MENTAL MAPS

MENTAL MAPS

SECOND EDITION

Peter Gould
and
Rodney White

London and New York

First published 1974 by Penguin Books
Second edition, 1986
Reprinted by Allen & Unwin Inc.

Reprinted 1992
by Routledge
11 New Fetter Lane, London EC4P 4EE

Simultaneously published in the USA and Canada
by Routledge
a division of Routledge, Chapman and Hall, Inc.
29 West 35th Street, New York, NY 10001

Set in 10/11pt Bembo by Computape (Pickering) Ltd, N. Yorkshire
Printed and bound in Great Britain by
Mackays of Chatham PLC, Chatham, Kent

*Library of Congress Cataloging-in-Publication Data
has been applied for.*

ISBN 0-415-08482-2

British Library Cataloguing in Publication Data

A catalogue record for this book is available from the British Library.

ISBN 0-415-08482-2

Still dedicated with affection and respect to
Halasz Bastya
for inspiration and understanding

Preface to the second edition

It is now nearly 15 years since *Mental maps* was written. In that time significant changes have occurred in the field of geography and in the world. In geography the enthusiasm for the quantitative revolution has quietened and become more subtle and reflective. The rate of innovation, particularly the application of statistics to spatial processes, is more measured and careful. We would be the first to admit that this is certainly a good thing, as many applications were developed in haste without proper respect for the implied assumptions. What remains of the 'revolution' is a much more careful commitment to *definition* and relationships, as well as a continuing concern for measurement and prediction, as hallmarks of the scientific method.

Changes in the world economy have created some pressure to produce a new edition. When we wrote the first drafts in 1969 and 1970 the world economy was expanding, and there was still some reason to believe that many of the developing countries were still developing. The worldwide recessions that have come and gone since the early 1970s now expose the context of our writing as being overly optimistic. However, the images that people hold of the components of their national space have changed more slowly than the economic reality. This lag effect has made all the more alarming the rate of rural–urban migration in those countries in which the economy is not even growing, let alone developing. This factor has made an understanding of the 'invisible landscapes' that people carry in their heads even more crucial. Indeed, some people concerned with predicting major migrational trends have noted that these images may well provide the best predictors of all. If the economic signals that are being transmitted from overcrowded cities – repositories of a growing number of unemployed in the Third World – are being filtered out by people's preconceptions, then the problem is serious indeed.

We are grateful for the many comments we received from colleagues and others who read the first edition. Criticism focused on the viability of the ranking of spatial preferences and the representativeness of the samples of respondents. Clarification of these and other points is included in this edition. The passage of time has made a detailed introduction to principal components analysis less necessary. Accordingly, the second chapter of the first edition has been shortened and placed in an appendix.

Peter Gould
University Park
Pennsylvania

Rodney White
Toronto
Ontario

Acknowledgements

The authors thank the following individuals and organisations for permission to reproduce copyright material (numbers in parentheses refer to text illustrations):

American Geographical Society (1.2); D. Ley (1.5); F. Ladd (1.6); Figure 1.7a reprinted from 'Urban experimentation and urban sociology', in *Science, engineering and the city*, publication 1498 of the National Academy of Sciences, Washington, DC, 1967 (pp. 103–17, by courtesy of P. Orleans); F. V. Thierfeldt (1.8, 1.9); B. Urquhart and Doncaster Development Council (1.10); S. Dornič (1.11); the Editor, *The Economist* (1.13); the Editor, *Regional Studies* (2.4, 2.6, 2.7, 2.12, 2.14, 2.15); Stephen Morgan (2.8); Figures 3.1–4 reprinted from *Spatial organization* (Abler, Adams & Gould 1971) by permission of Prentice-Hall Inc., Englewood Cliffs, New Jersey; B. Goodey (3.5, 4.1, 4.6); R. Lloyd and R. Van Steenkiste (4.6–8); T. Lundén (4.18–21, 6.12); Figures 4.24–9 reproduced from P. Gould and D. Ola 1970, The perception of residential desirability in the western region of Nigeria, *Environment and Planning* 2, 73–87, by permission of Pion Ltd; the Editor, *Area* (5.1); the Editor, *World Politics* (5.2); T. Leinbach (6.4, 6.5); G. Törnqvist (6.2); Richard Eaton (6.5, 6.6).

Contents

1 The images of places

We link together our various perceptual spaces whose contents vary
from person to person and from time to time, as parts of one public
spacio-temporal order. . . .

D. Hawkins, *The language of nature*

WHERE WOULD YOU REALLY LIKE TO LIVE?

Suppose you were suddenly given the chance to choose where you
would like to live – an entirely free choice that you could make quite
independently of the usual constraints of income or job availability.
Where would you choose to go? Probably, if you really *had* to make
such a choice, you would be assailed by images of faraway places, of
different climates and different landscapes, and by your personal
feelings towards cultures other than your own. You would also
become very sensitive to the affection of old friends, and the familiarity
of your present surroundings. You would be aware of the pull between
'here' and 'there' as places in which to live. The one offers the known
and valued, the other offers the unknown, the exciting and an escape
from the shortcomings of your present environment.

Differences between the attributes of 'here' and 'there' have always
been of great interest to human geographers, because it is precisely the
differences between places that generate movements of goods, people
and information. Such movements produce the traffic jams in cities,
crowds on summertime beaches, vast flows of letters and electronic
messages, and the great networks of pipelines that criss-cross nation
after nation. To analyse such complex patterns produced by man on
the surface of the Earth, geographers are increasingly looking at
questions of *relative* location – questions which consider places not in
any absolute, old-fashioned, latitude and longitude sense, but in terms
of their costs and times to all other places with which they might
exchange goods, money, people and messages. We would like to make
these varied patterns more efficient and increasingly satisfying to
people that use them, or are affected directly or indirectly by them in
many ways. The fact that we cannot satisfy all our demands in one
place is the root cause of these gigantic and varied patterns of exchange.
In the geographer's language these patterns are known as *spatial
interaction.*

Our choice of a residence is one attempt to reduce the amount of
travelling we must do, and to minimise the movements of goods,

1

information and people that must occur to satisfy our wants. The choice is quite a complex one as it is very unlikely that we can choose a place where we can satisfy all our demands completely. For example, if we are choosing a home in an urban area we might wish to be very close to work, but few of us would choose to live over the shop or right next to the factory. In fact, suburbanisation is a deliberate attempt to put some distance between the home and the congestion of the urban core and industrial areas. Again, many families like to live fairly close to their children's schools, but few would volunteer to live within earshot of the playground. 'Near, but not too near' is an interesting example of a spatial decision which is indicative of the nature of a whole range of problems to which geographers and other spatial analysts have turned.

If we were given a completely free choice of where to live, not just around a given city, but anywhere in the country, or perhaps in the world, the problem might become too much for an individual to handle. Certainly it would severely test the normal amicable agreements of most families. In actual fact, most residential choices are highly constrained by the realities of the world around us. Financial constraints, limited job opportunities, family obligations, children's schooling, language barriers, immigration laws and sheer personal inertia generally reduce the problem to a manageable level.

However, the world is becoming increasingly mobile. In the United States the average family moves every three years. In poor countries the rate of rural-to-urban migration has reached an alarming level, while the educated members of such societies look upon foreign travel as an essential experience. In a country like Britain, many young people are looking far from their local areas for their first job or for a university; and many of the unemployed migrate to escape regional economic decline. The growth in the size of companies with branch plants has forced many people to consider new appointments in different areas. Large groups of immigrants arrive in the country (some highly skilled, some with no skills) and they have few preconceived ideas of the qualities of different cities and regions. The wealthy among the retired, or those in chronic ill-health, may move to the countryside, to the south coast, or live abroad once they are really free to choose.

Rural-to-urban migration, regional drifts to escape unemployment, international movements – whether of the Brain Drain type or of unskilled labour – are subjects of common interest to the whole range of social scientists. Geographers are especially concerned with the effect of *distance* on movement – that is the effect of the *location* of an individual *relative* to his proposed destination. In addition, modern geographical studies of *environmental perception* look at the ways in which people form images of other places, and how these images influence many decisions – including the one to move. For example, a positive image may reduce the effect of distance – just as you will travel further to visit a good friend than you will to visit a more casual acquaintance. In these situations the ideas of social proximity and physical distance interact to produce a particular response, be it a journey, a letter, or a phone call.

As geographers, we are particularly interested in the way that the perceived differences between various parts of the Earth's surface affect movements of many different kinds. We shall look at a series of studies in which research workers have tried to assess the ways in which people build up images of other places. From the geographical point of view we are most interested in the way in which distance between the person and the place affects the manner of construction and the final product of this image-building process. We will also make some suggestions as to how such images might affect migration and other forms of spatial behaviour, though subsequent work will have to clarify and test the complex nature of these relationships. Our basic strategy is to put people in a fairly free hypothetical situation where they are asked to rank their order of preference for a series of places in terms of residential desirability. From maps of their 'space preferences' we attempt to explain the ways in which 'mental maps' are related to the characteristics of the real world.

Suppose you consider some of these questions in personal terms. Where would you like to live? And how do you deal with this question? What kind of sorting, evaluating and discarding process goes on in your mind as you begin to weigh up and compare all the places you know and have heard about? Many people, especially children, start thinking about holiday areas, for in their minds these are often associated with pleasant times in places deliberately chosen to be relaxed and different. Sometimes these vacation areas are chosen as permanent places of residence, and in nearly every country in times of economic expansion we have seen a swelling migration to the sun and sea. People on the point of retirement often face the real question, as opposed to one that is hypothetical for many of us, and move to sunny Devon and Cornwall in England. In the United States the locational loadstones are Florida, California, Arizona and New Mexico, and many other countries have their retirement areas. In Europe, Spain and Portugal are drawing large numbers of retired foreigners, seeking warmth for creaking bones and a clear atmosphere for old lungs.

On the other hand, it may be that vacation spots are not for us all the year around. Moving to new places means losing old friends, and facing the uncertainty of making new ones. For families with children this may be particularly uncomfortable and worrying. We know a great deal about the places we live in now, and few of us deliberately increase our level of uncertainty unless the rewards look quite high. Some, of course, are footloose and fancy-free, and choosing a new place to live in poses few worries. Others are more hesitant, and find that they are really very happy in familiar and comfortable surroundings.

THE NATURE OF DIFFERENT PLACES

As we begin to evaluate places in terms of their residential desirability, we may consider a number of very different aspects about them. In the first place, the amount and type of information we have about different localities will vary considerably. Some places we know from first-hand experience, and our information is direct and immediate. Other places are little more than names, and if pressed we would find it difficult to say much about them at all. Even our sources of information are extremely varied. While we acquire some through personal travel, we also form mental images of places with the information we get from reading, radio, television, talking to other people and even from travel posters in railway stations and airports.

The local scenery is certainly one thing many of us weigh in the balance. A flat, drab landscape rimmed by a horizon of chimneys belching smoke is less likely to appeal to most of us than an area of rolling hills or even majestic mountains. Certainly the split between rural and urban scenery is something nearly everyone considers in an evaluation of this sort. Some people are definitely town–dwellers, and enjoy the sense of bustle and exchange that is always present in large towns and cities. Many who have lived and grown up in cities feel quite lost when transplanted to the countryside, and they long to get back to the noise and services that large urban areas provide. But there are others who dislike cities intensely, and who would do almost anything to keep out of them. For them no amount of urban variety can possibly compensate for the noise, stench and general frenetic pace that seems to characterise these intense nodes of human activity. When you were a child, perhaps you read Beatrix Potter's *The Tale of Johnny Town Mouse*, that delightful juxtaposition of the suave and knowledgeable city mouse and his simple country cousin, Willie. After an exchange of visits, both decide that their own landscape is infinitely preferable to the other's, and there are many people who feel exactly the same way.

But scenery is obviously not the only aspect of a place that we consider as we weigh up our own feelings about living there. Climate enters our evaluation too, and we have already noted the value that many people place upon an equable temperature and sunshine. In fact, for some people certain types of climate are absolutely essential for their comfort, health and even survival. Advanced countries like Denmark provide chronic bronchial patients with a respite in the sun of the Canary Islands during the long damp winter months; in the USA, states such as Arizona and New Mexico are seasonal havens for rich people who suffer from asthma and respiratory troubles in colder northern areas. Of course, to some extent we can and do modify climate by heating our houses and buildings in winter, and by cooling them in summer. Enormous quantities of coal, gas and oil are expended every year on climate modification through household heating and cooling, and the energy requirements of a city like New York in summer can put such a strain on electrical generating capacity that 'brown-outs' and breakdowns occur with ever greater frequency. Until the 1950s, Washington DC was considered a climatic hell-hole in the summer for members of the British Diplomatic Corps, but today air conditioning is a substitute for personal hardship allowances.

Landscape and climate are not the only considerations as we think about places in terms of where we would like to live. Some countries contain areas of great cultural and linguistic variety, and, consciously or unconsciously, people tend to stay where the language is familiar and cultural reactions are predictable. Many French-Canadians, for example, do not like living outside Quebec, and often families will try to move back when their children reach school age. Many Scots prefer to educate their children in the Scottish school and university system, which they feel has a much more distinguished reputation than that south of the border. In some countries, such as Britain and the USA, local authorities may control education, and the quality of schooling varies markedly from one place to another. The quality of the local schools is nearly always one of the major concerns of families with young children as they think about moving to a new place. More sensible countries, like Sweden and Switzerland, do their utmost to ensure equality of educational opportunity, and while there are some differences between urban and rural areas these are generally much less important than in Britain and the USA. Political attitudes may also vary and colour people's evaluations of places. Few New Englanders want to live in Alabama and Mississippi because they still believe that immoral and archaic attitudes to race are expressed in political decisions at both the State, and, particularly, the local level. Similarly, some cities in the USA have a reputation for political and social corruption that shapes the mental images that people have of them.

Scenery and climate, cultural and educational diversity, and the variation in political and social attitudes do not exhaust the list of things we might consider as we weigh up the pros and cons of places. If many people wrote down everything they consider in an evaluation, we would undoubtedly have a long list of many minor aspects that could be considered. But there would probably be one thing that would not appear explicitly on many lists, although some people unconsciously consider it as they judge the desirability of places. This is the factor of relative location, or *accessibility* that we mentioned earlier. It is one of those common concepts that are apparently obvious and easy to understand, until we actually try to pin them down and measure them.

RELATIVE ACCESSIBILITY AND POPULATION POTENTIAL

Relative accessibility, the 'getatableness' of a place in relation to others, is a property of a particular location that may well enter our evaluation of its residential attractiveness. Geographers have tried to measure accessibility in a variety of ways, and have had to face the question: accessible to what? Our concern here is not accessibility to natural resources or things like that, but accessibility to other people and the services they can and do provide. One easy way of measuring this slippery notion is simply to lay down a circle with a certain radius, say 30 kilometres, upon a population dot map, and then record at the centre of the circle the number of people captured within that distance. If we centre our circle on hundreds of locations over the map, we can use these values of people captured as 'spot heights' to draw a contour map whose surface shows that variation in accessibility, measured in terms of the number of people who can be reached within 30 kilometres at each point. This is a little crude, however, and geographers have devised a somewhat better measure. The measure takes into account *all* the people in the country, rather than just those within an arbitrary distance, but discounts the contributions of those who are far away from the point on the map whose relative accessibility is being evaluated. This overall measure is called the *population potential* of a place.

As a very simple example, suppose that each person on the island of Ghageria Leone is represented by a dot on the map (Fig. 1.1a), and we want to calculate our measure of accessibility, or population potential, at place A in the middle of the island. People close to A should contribute heavily to our access measure, while those far away should be weighted less. One way of discounting access to a population with distance is simply to count up the number of people in a series of rings drawn around the location, and then divide these totals by the number of kilometres away. For example, in the first ring there are four people, so we divide this value by one. In the next ring there are seven people, so we divide this amount by two. If we do this for each ring, and add all the values together, we find that the population potential at A is 17.65. On the other hand at B, near the coast of Ghageria Leone (Fig. 1.1b), the population potential is only 11.92. According to our simple measure, Place A has more overall accessibility than B. Notice how these values depend both upon the way the people are distributed, and the distance they are away from the point whose population potential is

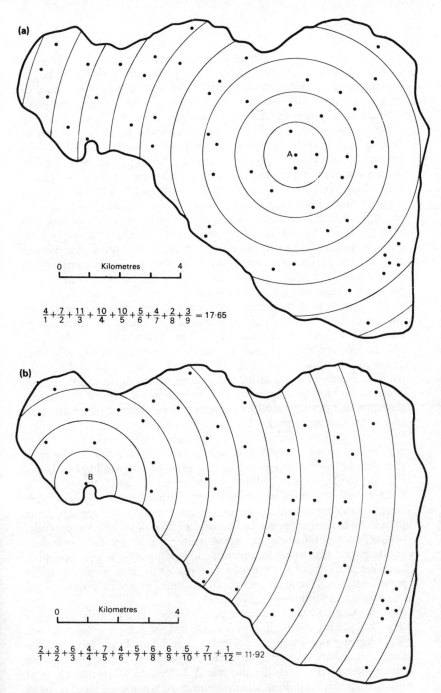

$$\frac{4}{1}+\frac{7}{2}+\frac{11}{3}+\frac{10}{4}+\frac{10}{5}+\frac{5}{6}+\frac{4}{7}+\frac{2}{8}+\frac{3}{9}=17\cdot65$$

$$\frac{2}{1}+\frac{3}{2}+\frac{6}{3}+\frac{4}{4}+\frac{7}{5}+\frac{4}{6}+\frac{5}{7}+\frac{6}{8}+\frac{6}{9}+\frac{5}{10}+\frac{7}{11}+\frac{1}{12}=11\cdot92$$

Figure 1.1 (a) Measuring the population potential at an interior point (A) in Ghageria Leone. (b) The population potential at a coastal point B.

being measured. Although geographers have devised more realistic and complicated measures of access, by weighting different populations and discounting their contribution with complex measures of distance, time and cost, the basic principle underlying the most difficult measures is the same as the one we have used here.

Of course, when people consider a particular place in terms of its overall access, they do not start cranking through a lot of tedious arithmetic to calculate precise measures. Most people have a fairly strong intuitive feel for the core of a country and its more peripheral areas. Many like to be close to 'the centre of things', and this means in an area with high population potential. In Britain (Fig. 1.2), for example, values have been calculated by computer for 90 locations all over the country, and these have been used as control points, or spot heights, to produce a smooth surface of population potential. Notice that the intervals between the contour lines are geometric, rather than even arithmetic steps, in order to display the marked contrasts in population potential across the country.

The highest peak on the surface is immediately around London and the South-East, with a ridge of accessibility through the Midlands to another peak between Leeds and Liverpool. While many people do not like the idea of living in London itself, they want to be reasonably close to the metropolitan area so that they can get 'up to Town' if they want to. The South-East has also been an area of strong in-migration for decades, so that people have been reinforcing the very population potential that draws them like a magnet. Now the area is becoming so congested that attempts are being made to halt the 'drift to the South'. Notice how a new link, such as the Channel Tunnel, will send a jolt of continental population potential across the water to reinforce the areas that already have the highest levels. In terms of accessibility to a larger Europe, the South-East will become an even more attractive place. To the South-West a third peak appears over Bristol, but the gradients are steep everywhere, and much of England and Wales appears quite peripheral to the core areas. In Scotland a hump of population potential occurs between Glasgow and Edinburgh, but it is so small that it does not show on this map. In general, the surface reflects most people's intuitive ideas about the most accessible area in Britain and the more peripheral, outlying areas such as Devon and Cornwall, Wales and the Lowlands and Highlands of Scotland.

In the United States a similar map of accessibility has been drawn (Fig. 1.3), with very high peaks and ridges between Boston and Washington, Pittsburgh and Chicago, and San Francisco and San Diego. It is hardly an accident that these are precisely the areas that many predict will merge by the year 2000 into the Boswash, Chipitts and Sansan megalopolises, the first aspects of what some have referred to as ecumeniopolis, or a worldwide urban system. Many things in

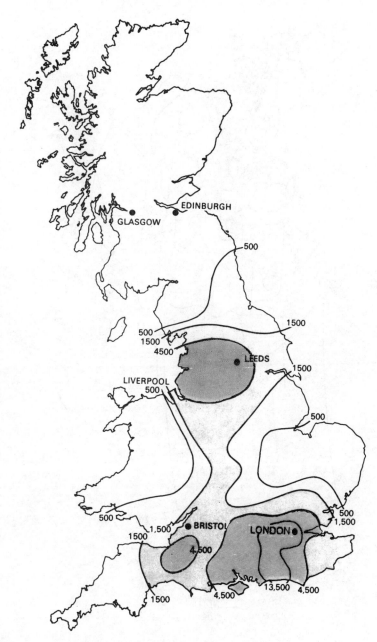

Figure 1.2 The population potential surface of Britain.

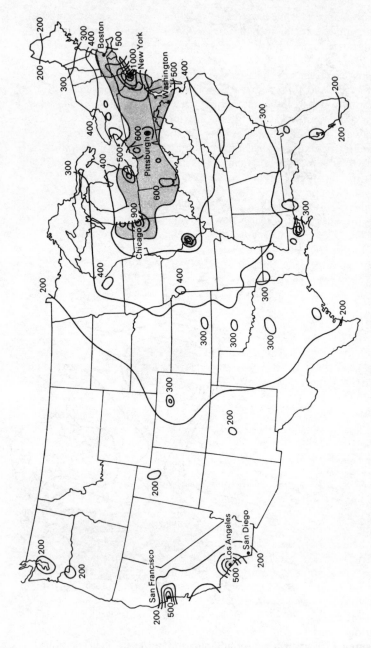

Figure 1.3 The population potential surface of the United States (adapted from W. Warntz).

American society vary in an extremely regular manner with population potential, and the areas with high values have traditionally been places of in-migration as people have demonstrated their preferences with their feet.

We must not over-emphasise this business of accessibility. It is only one of many things that people consider, consciously or unconsciously, as they evaluate where they would like to live. Today the stress of living in areas of high population potential may even drive people away from the peaks, rather in the same way that heavily congested traffic in the centre of the city may make motorists avoid the down-town areas. But while many of us like to get away from it all for a short time during a holiday, most of us hate the idea of living in a really isolated place for a long period. In our archaic and almost medieval prison systems we even punish men and women by the extreme form of social isolation – solitary confinement. We may not want to live in a penthouse overlooking Piccadilly Circus or Times Square, but many feel it is rather nice to be reasonably close to things. And this is precisely what population potential measures – the closeness of people to people.

PERCEPTION OF THE ENVIRONMENT

Although we knew very little about people's preferences for places until the 1960s, and while much remains to be done, a number of people have thought about the images that men have of their local environments. Usually these environments are urban areas, and interest has focused on the way in which people perceive certain landmarks, routes, boundaries and neighbourhoods. One of the first people to comment upon such things was Charles Trowbridge in 1913, when he noted that some people in a city always seemed to have a good sense of orientation, while others '. . . . are usually subject to confusion as to direction when emerging from theaters, subways, etc.' Some people, he thought, had informal, imaginary maps in their heads centred upon the locations of their homes. They were able to move around the urban landscape as long as they remained on familiar ground, but they quickly become disoriented in unfamiliar areas. Others appeared to be egocentric, and see directions in relation to their own position at the moment. These people seemed to be able to navigate much more surely, and Trowbridge even went so far as to recommend directional training for children in schools.

For some reason few people followed up the early anecdotal leads of Trowbridge, but in the 1950s Kevin Lynch raised the question of environmental perception once again with his book, *The Image of the City*. By asking a group of people about their feelings for prominent landmarks in Boston, Jersey City and Los Angeles, and questioning them about major routes and areas they used in driving around, he was able to build up a general image of the city that pulled out the basic elements of the urban landscape. Interestingly, Lynch's concern for the information people have about the city of Boston has been translated into practical planning terms in Birmingham. Brian Goodey, with the help of the *Birmingham Post*, asked people to co-operate in a study investigating their perception of the city centre (*Birmingham Post* 1971, Goodey 1971). He asked readers to send in maps that conveyed the major impression they had of the area. Accuracy for its own sake was not required, and no published maps were to be used by readers in drawing the sketches. What was wanted was a quick, unaided impression to give the basic pieces of information that people had in their heads, and which they used in moving around the centre of Birmingham. The response was very large; people seemed to like the idea of helping planners and being involved in some small way with the planning process going on in *their* town. By combining many hundreds of responses (Fig. 1.4), the planners were able to build up a weighted mental image that seemed to emphasise a marked preference for things at a human scale. Some of the tall, skyline features were not nearly as prominent as expected, while others such as the Cathedral Yard and the Bull Ring were singled out as desirable 'oases' in the bustling urban scene. Many specific shops were mentioned because of their street-level interest, while other areas were virtual blanks. Such a map, and the large number of comments about the area, proved very valuable to planners trying to think through the future appearance of this city.

Cities are not always pleasant places to live in, and the information that goes into building a mental image of a particular area may reflect much more than just the knowledge of landmarks and routes. In the USA many urban places are highly stressful, even dangerous, environments, and we can think of invisible surfaces lying over these areas, whose peaks represent places of high psychic stress, while the valleys are safer channels through the urban jungle. In a predominantly black area of northern Philadelphia, for example, David Ley (1972) mapped a large sample of the local people's fears as an environmental stress surface (Fig. 1.5). There is an invisible, mental topography of psychic stress in this neighbourhood, where the peaks are places to be avoided, while the lower areas and valleys are areas of greater safety. The peaks generally coincide with the headquarters of gangs close to the centre of their turfs, areas of abandoned building, and places where drugs are peddled. For an adult or child living in the area, information about this invisible topography literally lets them survive in a physically dangerous environment that is solely the creation of man.

People's information about a particular area in one of the USA's

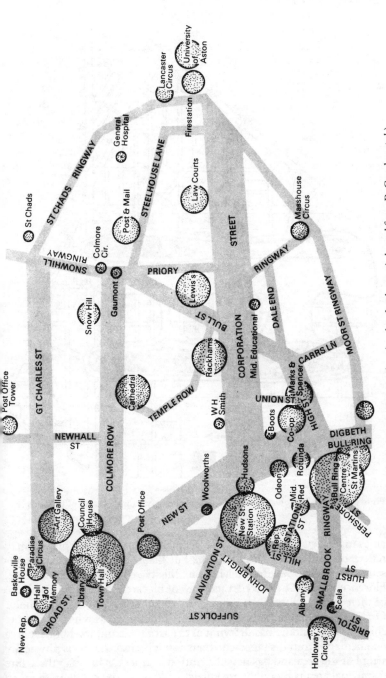

Figure 1.4 The people's preferences for features in Birmingham's urban landscape (adapted from B. Goodey et al.).

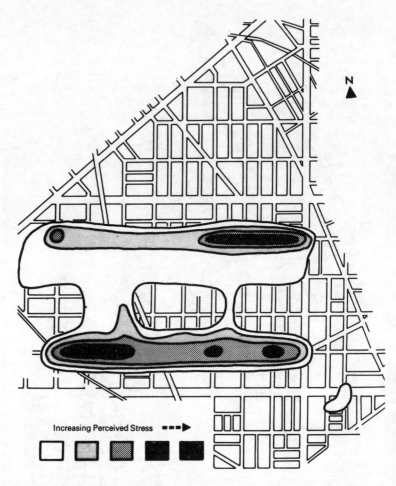

Increasing Perceived Stress ■■■▶

Figure 1.5 The perceived environmental stress surface for a portion of Philadelphia.

cities may vary considerably, and the mental images they build up may reflect not only their surroundings but many other aspects of themselves and their lives. In the Mission Hill area of Boston, for example, Florence Ladd (1967) asked a number of black children to draw a map of their area, and then she tape-recorded her conversation with them. On Dave's map (Fig. 1.6a), the Mission Hill project is where the white children live, and he has drawn it as the largest, completely blank area on his map. From his taped conversation it is clear that he is physically afraid of the area and has never ventured near it. On his map the white residential area is literally *terra incognita*, while all the detail on the map is immediately around his home and school on the other side of Parker Street. Ernest also puts in Parker Street dividing his area from the white Mission Hill project (Fig. 1.6b), and uses about a quarter of his sheet of paper to emphasise, quite unconsciously, the width of this psychological barrier. Both of these boys, going to the local neighbourhood schools, have never ventured across this barrier to the unknown area beyond. However, another black youth, Ralph, who attends the well-known Boston Latin School, draws a completely different map (Fig. 1.6c). The white Mission Hill project is greatly reduced in scale, and he puts in five educational institutions in the area, indicative of his perception of education as an escape route from the segregated life he leads. Similar pieces of information are given quite different emphasis in the mental images these boys have, and the patterns of information even begin to define their neighbourhoods. Dave and Ernest obviously feel at home only in a quite restricted area which they know well, while Ralph has a much wider view and allocates his information evenly across the map.

The concept of a neighbourhood is an important mental image, both to the town planner and the rest of us who are subject to planning. We have much evidence today, from many of the world's cities, that breaking up a cohesive neighbourhood can have many detrimental social and psychological effects. The question of measuring this social space, which has such an important degree of familiarity for a

(a)

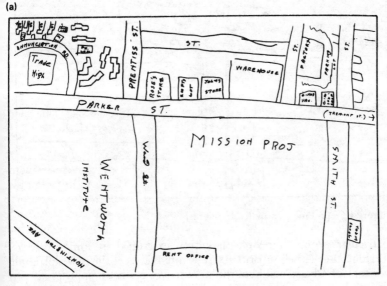

Figure 1.6 (a) Dave's map.

(b)

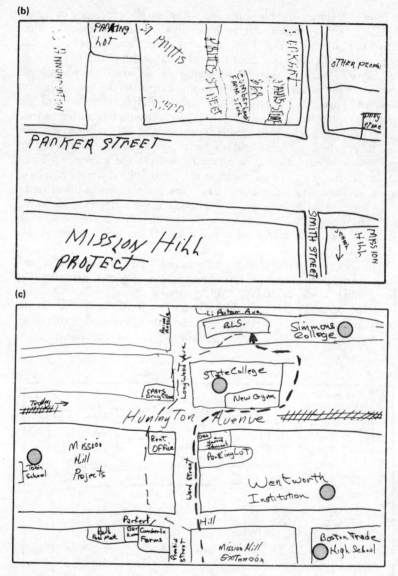

Figure 1.6 (b) Ernest's map. (c) Ralph's map.

particular group of people, has been examined in some detail by Terrence Lee (1963) in Britain. He wanted to see if the town-planning concept of a basic neighbourhood unit was really an appropriate one for modern urban living. What he discovered was that social space and physical space are so tightly linked that most people simply do not distinguish between the two. On the average, people tend to define their neighbourhoods as an area whose size seems to be quite independent of the density of the people living in it. Neighbourhoods in outer middle-class suburbs and high-density slums are perceived as about the same size. In other words, people do not think of their neighbourhoods in terms of the number of people, as planners often do, but only as a comfortable and familiar space around them.

On the other hand, the wider knowledge that a group of people possess of their city may vary very markedly with both their social class and location. In Los Angeles, for example, Peter Orleans (1967) questioned a wide range of groups in the city, and from their responses constructed composite maps showing how the intensity of their knowledge varied over the urban space. Upper class, white respondents from Westwood had a very rich and detailed knowledge of the sprawling city and the wide and interesting areas around it (Fig. 1.7a), while black residents in Avalon near Watts had a much more restricted view (Fig. 1.7b). For the latter, only the main streets leading to the city centre were prominent, and other districts were vaguely 'out-there-somewhere', with no interstitial information to connect them with the area of detailed knowledge. Most distressing of all was the viewpoint of a small Spanish-speaking minority in the neighbourhood of Boyle Heights (Fig. 1.7c). This collective map includes only the immediate area, the City Hall and, pathetically, the bus depot – the major entrance and exit to their tiny urban world.

The images that people have of places are not confined simply to local areas within the city. We all have opinions about various parts of the country we live in, and we tend to view the people who live around us in decidedly different ways. Some of these viewpoints have been expressed in rather humorous ways that contain, as humour so often does, that sharp edge of truth that makes us inwardly agree even as we smile at ourselves and others. Two of the most famous examples are Daniel Wallingford's maps of the New Yorker's and Bostonian's views of the United States. As far as the New Yorker is concerned (Fig. 1.8) the boroughs of Manhattan Island and Brooklyn are perceived as most of New York State, while the northern borough of the Bronx is so far away that to all intents and purposes it is close to Albany and Lake George in the Adirondaks. Connecticut is somewhere to the east, but it is mentally blurred with a place called Boston – obviously a mere village in the crook of Cape Cod. Westwards the map becomes more and more distorted and hopelessly inaccurate. States exchange names and wander to outlandish locations; California has been partitioned between San Francisco and Hollywood (after all, what else is there?), and Florida has swollen to assume its proper proportion as the holiday place for New York just south of Staten Island.

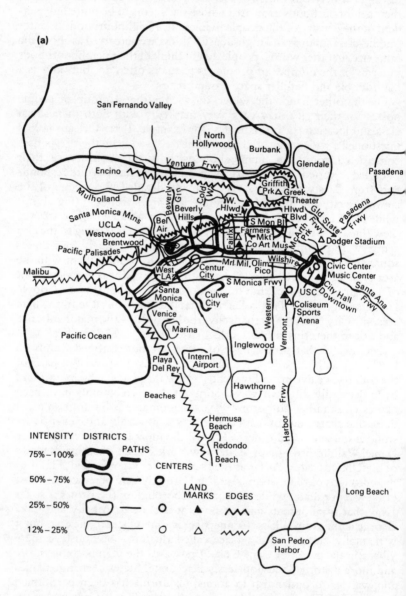

Figure 1.7 (a) Los Angeles perceived through the eyes of upper-middle-class whites in Westwood. (b) Los Angeles through the eyes of black residents in Avalon. (c) Los Angeles through the eyes of Spanish-speaking residents in Boyle Heights.

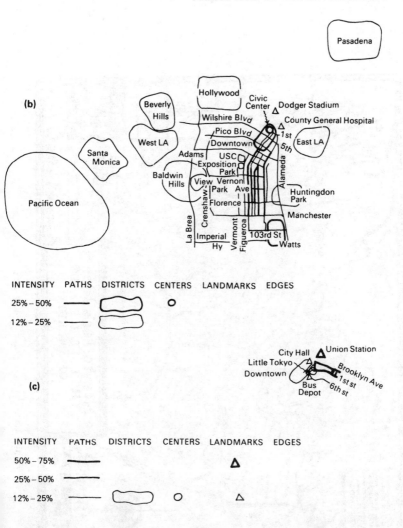

But from Boston (Fig. 1.9) things appear in a somewhat different light, as the great primate city in the centre of New England – the major region of the United States – assumes its proper cartographic proportions. Brash, somewhat *nouveau*, towns like New York and Washington are put in their proper places in the hinterland, for it is clear that they are merely stopping points on the fringe of the West Prairies, where presumably other towns exist. Cape Cod is a magnificent sheltering arm, beckoning commerce and culture from the four corners of the Earth. We smile and yet deep down we recognise ourselves in the New Yorker and the Bostonian, with their hopelessly

Figure 1.8 The New Yorker's view of the United States.

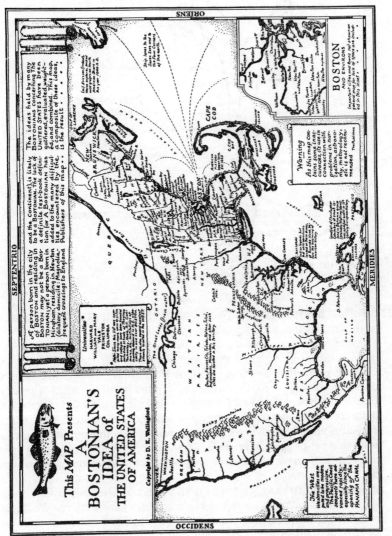

Figure 1.9 The Bostonian's view of the United States.

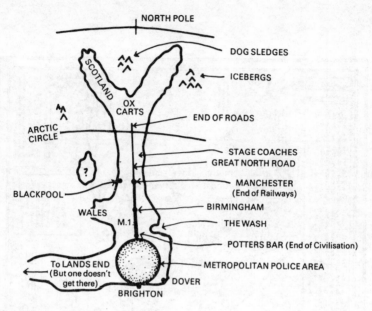

Figure 1.10 How Londoners see the North – at least, according to the Doncaster and District Development Council.

parochial but steadfastly proud views of their own towns at the centre of things and the rest just 'somewhere out there'.

Sometimes such distorted and humorous views can even serve useful ends. In England, the Doncaster and District Development Council published Ye Newe Map of Britain (Fig. 1.10), as a way of drawing attention to the manner in which many southerners in England misperceive the North. The council wanted to attract more industry and jobs to the Doncaster area, and realised that many decisions affecting business locations were made in London. The idea was that if people have their legs pulled in this way they will re-examine their spatial biases and make more informed decisions in the future.

We all tend to exaggerate our own town and area, and usually look askance at someone who has never heard of where we live. We know more about the areas close to us, and they tend to become much more important than others about which we know little. Our emotional involvement with other places changes quite markedly with our subjective estimates of how far places are away from us. For example Stanislav Dornič (1967), following the work of Gösta Ekman and Oswald Bratfisch (1965) at Stockholm University, has shown how people's emotional involvement, the degree to which they really care

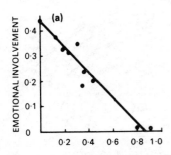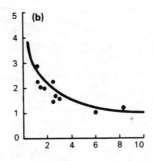

Figure 1.11 (a) The degree of emotional involvement related to distance in logarithmic co-ordinates. (b) Emotional involvement related to distance in standard co-ordinates. (Adapted from S. Dornič.)

about what happens at a place, falls off in an extremely regular way with their estimates of how far the places are away from them (Fig. 1.11a). From the slope of the line in logarithmic co-ordinates we can see that, on average, people's interest falls off with roughly the square root of the distance (actually 0.47, rather than exactly 0.50). When we consider this very precise and often replicated relationship in ordinary co-ordinates (Fig. 1.11b), we can see how people's emotional involvement with other places falls off very steeply with distance, and then more or less levels out beyond a certain distance. We can even make a map projection centred on our home in which places close to us loom very large, but tend to shrink more and more as we move farther and farther away. Such a map projection has been drawn by Torsten Hägerstrand (1953) at Lund University, and while it was originally constructed for other purposes (plotting migration data in Sweden), it may well be that it conveys quite accurately the general mental picture that people have of places around them. Certainly it is naïvely obvious to one of us that the village of Van Hornesville, which obviously needs no further identification, is at the very centre of things (Fig. 1.12). On our logarithmic projection the area of central New York State assumes its rightful importance, after centuries of distortion and neglect by cartographers whose ignorance one can only lament. Notice how some subcentres such as New York, Washington, London and Moscow, are simply satellites in the backwaters of civilisation at the world's edge.

In England, the village of Edgcote plays a similar central role in C. Northcote Parkinson's devastating geographical analysis of social class and movement (Fig. 1.13). People tend to be As (aristocratic, anarchic or artistic), or Bs (bourgeois, bureaucratic or benevolent), and their social ladders from desperation to privilege have distinct geographic expression. The two systems are divided by a line from Hastings through London, splitting the West End and the City, the Commons

23

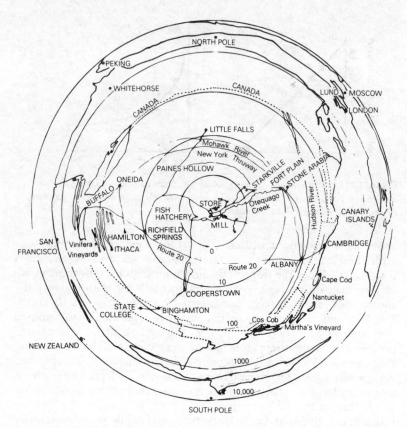

Figure 1.12   Van Hornesville as the world's nodal point.

and the Lords, and then on to Mad Freshfield Wharf near Liverpool. A family in the B system may climb from Manchester or Sheffield, via business and Cambridge, to the privileged area of Essex, Surrey or Kent. At this point the grandsons of the shrewd industrialist who started the cycle may gain commissions in the Guards, cross to the A system and through extravagance, gambling, drink and sex begin the cycle down again to the area of desperation at Liverpool. Within the A cycle, passive families in Shropshire ascend via Oxford to Whitehall or Portland Place, and then to the heart of privilege in Berkshire. It is impossible to do full justice to Professor Parkinson's penetrating analysis here, but English readers particularly are urged to consult *The Economist* of 25 March 1967 (Parkinson 1967).

Again we may smile, but even as we do so we have to acknowledge that this map is a splendid example of highly localised spatial perception. Most of us are prepared to laugh at our own provincialism even as

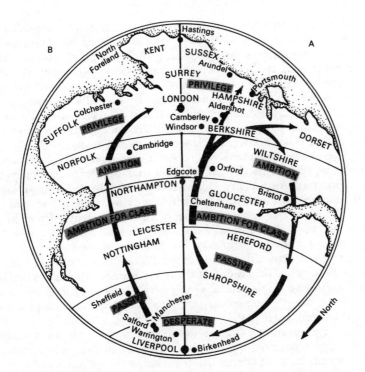

Figure 1.13 England projected into social space.

we flaunt it, and we accept that others may have different views about their areas. In the USA there are many Easterners who have a mental image of Iowa as the epitome of rural backwardness, with people living in small towns dotted through flat cornfields. The little old lady from Dubuque has become a stock character. Iowans, of course, know better, and smile pityingly at such views that are ignorant of the state's deeply rooted cultural wealth and its beautiful landscape. Few Iowans would willingly trade their heritage for the frenetic pace, rudeness and silly posturing of the eastern seaboard. Their views condition their decision to move or stay, even as the Easterner's perception of the Midwest influences his willingness to leave behind places and people that are familiar and comfortable.

THE GEOGRAPHER AND SPATIAL PERCEPTION

We are slowly realising that people's perception of places is one of the things we must consider as we try to understand the pattern of man's work on the face of the Earth. The geographer's concern for the way people perceive their environment is nothing new, but his approach to the problem changed rapidly as geography focused on problems of applied interest and as interdisciplinary linkages grew. Some of the most penetrating studies have come from historical geographers, who have tried to convey the perception and evaluation of new environments by pioneer explorers and settlers. By carefully analysing contemporary sources, Ronald Heathcote (1965) has recreated the views of early Australian pioneer farmers and herders along the Queensland and New South Wales border, and James Wreford-Watson (1969, 1970–1, 1971) has shown how men's perceptions of the great prairies of Canada have changed dramatically over the centuries. Similarly, the comparative American–Australian work of Donald Meinig (1962) has illuminated the way attitudes to rather similar environments are shaped by closely aligned but distinctly different cultural systems. These ideas have also been incorporated into a geographical learning game for high-school students, who take on the roles of pioneer farmers in western Kansas towards the end of the 19th century and try to outwit the environment. This is one of the many learning games devised by the imaginative High School Geography Project in the USA in an attempt to bring contemporary geographical insights and approaches into the school curriculum.

But the geographer's concern for perceptual questions is not simply confined to reconstructing the views men had of past environments. Many of the human patterns we see on the landscape today are a result of men making locational decisions based on information that has come through a perceptual filter. For example, a great deal of geographical research has been conducted by Gilbert White and his students on the perception of environmental hazards, in an attempt to understand why some people make apparently irrational decisions in the face of catastrophic natural events such as floods, coastal storms and earthquakes. Although these appear in a random and quite unpredictable fashion, we often know a great deal about the frequency with which such events occur. We can often predict, with a very high level of probability of being right, that a flood or storm of a certain magnitude will occur within a particular time period. Unfortunately, while a prediction that a hazardous natural event will occur with a high degree of probability is usually very accurate, we are still incapable of predicting exactly when it will occur! Before the year 2000, for example, it is highly probable that the Thames will experience an extremely rare high tide that could flood the lower areas of the city, pour into the Underground, and destroy much of London's famous transport system. This event is perceived as being very unlikely tomorrow (which it is), and so for a long time London did nothing to prepare for this unlikely, but in the long run almost certain event. Today, a huge flood barrier is in place, ready to stop the waters before they rise too high. Similarly, it is highly probable that California will experience a frightful earthquake before the turn of the next century. But it is clear that the collective memory of the cataclysmic San Francisco earthquake in 1906 has decayed over time, and today few buildings are built to the structural standards required by laws enacted immediately after that devastating event. A tremor in Los Angeles in 1970, in which a whole hospital collapsed like a child's sand castle, reinforced people's memories, but today it seems that their perception of such a hazard is blurring the sharp jolt of reinforcing information they received only a few years ago.

When geographers use the term 'environment' today, they do not think simply of the physical environment, for they have greatly enlarged their definition to take in the man-made and social environments that are usually of much greater importance in human affairs. How men perceive their physical and social environment is a crucial question for the contemporary human geographer. It is also important for the way it directs the geographer's attention to other areas of the human sciences in which environmental questions are rapidly emerging. One of the most penetrating human scientists, Herbert Simon (1969), has noted that it may be useful to think of human behaviour as being quite simple, but that most people live in very complex physical, man-made and social environments so their actual behaviour *appears* extremely complicated. He draws an analogy with an ant moving over a beach ribbed with waves of sand: a map of the ant's path might appear very complex, although the ant's behaviour was directed at achieving her simple goal of getting back to her nest. The crooked path we see reflects the complicated wavy surface or environment in which her simple task is carried out, rather than truly complex behaviour on her part. Of course, this is only an analogy, and given Simon's deep concern for understanding the human condition he would be the first to deny that men are simply ants. His approach to understanding human behaviour is a most intriguing one, however, and virtually reverses traditional viewpoints in which the extreme complexity of man's behaviour is seen to lie within him. The assumption that human behaviour is simple may be very fruitful, just because it directs our attention to the environments in which people's lives are embedded. It has already led to some models of man that are extremely thought-provoking and make good intuitive sense.

The importance of the environment is also emphasised in the controversial work of the psychologist B. Skinner (1971), whose ideas on operant conditioning were formulated after years of meticulous experiments with animals. His basic thesis is that behaviour can be altered by controlling the environment in which the behaviour takes place. Since we are already controlling human behaviour, whether we realise it or not, we should recognise this explicitly so that any alterations in behaviour by manipulating the environment can be directed to useful social ends. His proposals have generated very violent reactions, for his ideas evoke the spectre of 'Big Brother', and many are worried about the control of the controllers. Some have pointed out the basic circularity of his argument, for the controllers themselves will have been conditioned by a controlled environment since childhood, and will continue to live their lives under such conditions. Nevertheless, his explicit recognition that the human environment is a determining force in man's behaviour is going to have a lasting impact upon human affairs.

But there is one thing we must recognise when we start thinking about man-environment relationships – again, using the term environment in the broadest sense. Human behaviour is affected only by that portion of the environment that is actually perceived. We cannot absorb and retain the virtually infinite amount of information that impinges upon us daily. Rather, we devise perceptual filters that screen out most information in a highly selective fashion. Our memory, far from holding every sensory impression from our environment, selects and retains only a small portion. Our views of the world, and about people and places in it, are formed from a highly filtered set of impressions, and our images are strongly affected by the information we receive through our filters. This is why filter control is so crucial, both on a personal, individual basis and on a larger governmental scale. For example, good education should help to open up a child's filter on the world, so that his opinions and impressions are not based upon a highly selected and biased set of information that can so easily lead to nationalism, racial prejudice and the sort of crippling snobbery that still characterises so much of English life today.

At the national level, filter control is just as damaging, because the manipulation of vital sources of information can alter the whole outlook of a nation. In France, for example, the government's control of television has resulted in few programmes on other European cultures, and virtually no discussions on questions of European unity. Since television is a major source of information for most French people, the way its programmes are filtered by government policy greatly affects the people's perception of the place of France in the larger European family. On the other hand, the lack of filter control in American television has resulted in the wholesale pollution of the information channels with programmes whose purpose is to absorb time by purveying vapid commercialism, extreme violence and highly selected and exaggerated coverage of current events. The geographer, Brian Berry, for example, has noted the way in which the image of today's city is largely the product of information passing through the television filter, and he ascribes the tidal wave of white movement to the suburbs at least in part to people's perception of 'the city' as an evil, violent and race-racked place (Berry 1970).

The perception that people have of places, and the mental images

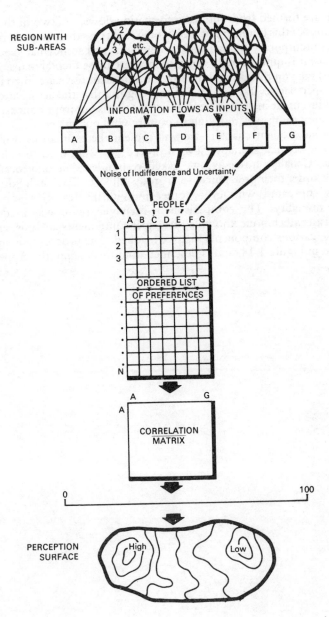

Figure 1.14 Constructing a preference surface: from the raw information to the final surface of the mental map.

that are formed from filtered information flows, are two of the major themes of this book. But it is one thing to discuss these important facets of contemporary human society in general and somewhat speculative terms; it is quite another to try to measure these mental maps. Yet we must take on this difficult task of measurement if we are to go beyond the speculative stage, even as we recognise that the act of attempted measurement may itself be a filter that screens out important aspects of the very thing we are trying to capture with our rulers.

In subsequent chapters we will present the results of a variety of studies of mental maps. Several procedures were used to collect the data. Usually, however, we asked respondents in one location to rank-order their preferences for a group of places. Each individual's list was compared with every other individual's list in a search for communality. The comparison of each viewpoint was made by a statistical technique known as principal components analysis, and it is these various components that we mapped. The procedure is summarised in Figure 1.14, and is described in more detail in the Appendix.

2 Images of Britain

When I am living in the Midlands
    That are sodden and unkind . . .
And the great hills of the South Country
    Come back into my mind.

Hilaire Belloc, *The south country*

For many people the question, 'If you had a free choice, where would you like to live?' has a somewhat speculative air about it. Many of us are fairly content to live where we are at the moment, and even if we had a completely open choice, we might not pack up immediately and rush to locations high on our preference list. The thought of moving conjures up horrendous pictures of removal vans and family chaos, and if we have children it means taking them out of familiar schools and making them go through another process of adjustment to unfamiliar places and faces. For most families a move to a new location is not something undertaken very lightly.

Yet there are times in our lives when the thought of moving does not seem difficult at all. On the contrary we come to certain stages when we feel we would like to break away from familiar routines and strike out afresh. One of these periods is when we are in our late teens and are about to leave school. This means breaking a routine we have been following for 10 or 12 years; a routine which may have seemed increasingly irritating to us as we grew older, even as it gave our lives considerable security and predictability. Suddenly the end of school is in sight, and the certainty of our well established routine is replaced by the uncertainty of what we are going to do next. There are few who do not feel at least a twinge of apprehension. At the same time parental ties may be chafing a bit, and conforming to home routines may be increasingly irksome. The thought of moving away, and being responsible for one's own life, can appear very exciting and attractive at this point.

For school leavers the question, 'Where would you like to live?' not only has an air of considerable reality about it, but it may be a question that young adults frequently raise themselves at this critical stage of their lives. It was for this reason that several hundreds of school leavers were questioned during the summer of 1967 at 23 schools scattered all over England, Wales and Scotland (Gould & White 1968). The young men and women were given maps of Britain divided into the 92 counties (as there were then; post-1974 county names are given in brackets) and were then asked to record directly upon the map their

own, quite personal preferences for residence. Thus a group of school
leavers at each school provided a matrix of rank-ordered preferences,
and the following mental maps were constructed from these sources of
information in the manner we described in the last chapter.

THE BRISTOL VIEWPOINT

The mental map of the school leavers at Bristol (Fig. 2.1), is quite
typical of viewpoints in the South of England. There is a strong
preference for the area immediately around Bristol and Somerset,
although the surface actually rises in the South-West to a peak of
desirability over Devon and Cornwall. It is possible that the idyllic
view of these two counties is generated by frequent holiday visits from
nearby Bristol, although, as we shall see soon, the South-West of
England has an extraordinarily bright image in nearly all people's
minds. In fact all along the south coast there is a high ridge on the
mental map, but in two areas the gradient of the surface to the north is
steep. Centred over the London area is a sinkhole of relative dislike,
and it is clear that this metropolitan area holds no particular charm for a
group of young Bristolians. The second steep gradient lies to the
north-west of Bristol, between England and Wales. It is one of the
steepest gradients on any mental map we have ever examined, for the
surface falls from 94 to 21 over a distance of less than 80 km as the crow
flies. For this group the northwards view over the Bristol Channel is a
view to an alien land, although Pembrokeshire and Caernarvonshire
(parts of Dyfed) escape the blanket condemnation. We must speculate a
bit at this point, but it seems unlikely that many have visited Monmouthshire (Gwent) and Brecknockshire (part of Powys) across the
water, because in 1966 travel to these areas was long and very
circuitous. Much of the dislike seems to come from a lack of information about south and central Wales, but perhaps the Severn Bridge
has since increased the flow of information about Wales, and hence
eroded the sharp mental escarpment has been eroded somewhat.
    As the view moves northwards the trend on the perception surface is
nearly always down, although the gradual decline is not at all regular.
Running through East Anglian counties such as Cambridge and
Norfolk, and through the beautiful border counties of Hereford
(Hereford and Worcester) and Shropshire, are ridges of relatively high
desirability, and these form the eastern and western walls of a mental
cirque lying over the Midlands. Indeed, the 30 isopercept slices right
through the Midlands, separating Britain into an area of rugged
desirability and an undulating northern plain of dislike. Only when
Westmorland and Cumberland (parts of Cumbria) are reached does the
surface rise sharply from the floor of the plain as though an intruded

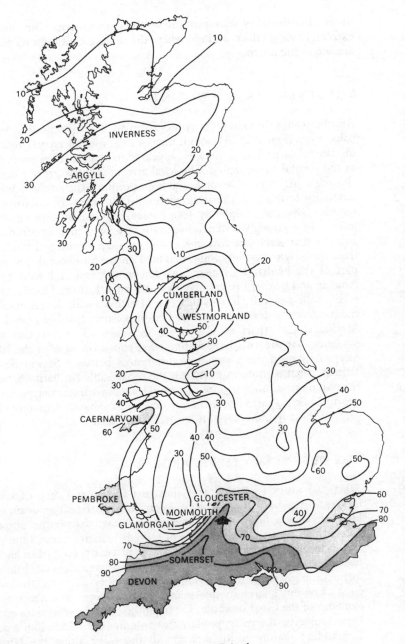

Figure 2.1 The mental map of school leavers at Bristol.

dome of desirability were pushing up the Lake District. But this is an extremely local effect, and elsewhere the surface continues its gradual decline to the north.

THE SEVENOAKS VIEWPOINT

Another southern viewpoint comes from an all-girls school in Seven-oaks, Kent (Fig. 2.2). Although many of the overall aspects resemble the Bristol mental map, the girls appeared to accentuate many features of the mental topography. The local area of Kent is greatly desired, although once again Devon is the high point on the surface. What is especially striking is the deep pit of dislike over the metropolitan region. Given a completely free choice, obviously nothing would induce these young women to live in the city, or even in its commuting suburbs that reach like tentacles across the map. So deep is the dislike from the Sevenoaks viewpoint, that the London Sinkhole almost forms part of the Midland Cirque, bounded to the east and west by the Anglian and Border Prongs. To the north only the Lake District Dome relieves the generally low-lying plain of dislike, with the exception of the moderately desirable areas of Pembrokeshire (part of Dyfed) and Merionethshire (part of Gwynedd) in Wales. There is also some evidence that the girls have a somewhat romantic view of the Isle of Skye. Although there are some differences between Sevenoaks and Bristol, what is more important is the remarkable similarity between the two maps of these groups of school leavers with southern locations nearly 240 km apart. The main features, and the basic mental topography, of these mental maps are virtually identical.

THE ABERYSTWYTH VIEWPOINT

When we move to Wales, and construct the mental map of school leavers at Aberystwyth (Fig. 2.3), a somewhat different view appears. First, there is a high peak of desirability centred over the point of perception in Cardiganshire (part of Dyfed), which indicates the very strong feelings for the local area. No other place is perceived so highly, with the exception of Kent, which joins with Sussex and Hampshire to form the high point on the ridge of desirability all along the south coast. London shares in this general approbation, and there is no evidence of the usual sinkhole. Far from the city, the school leavers of Wales perceive the metropolitan area as quite desirable. The second outstanding feature is the strength of the ridge along the border, forming the west wall of the barely accentuated hollow over the Midlands. This sharply etched feature not only includes the traditional

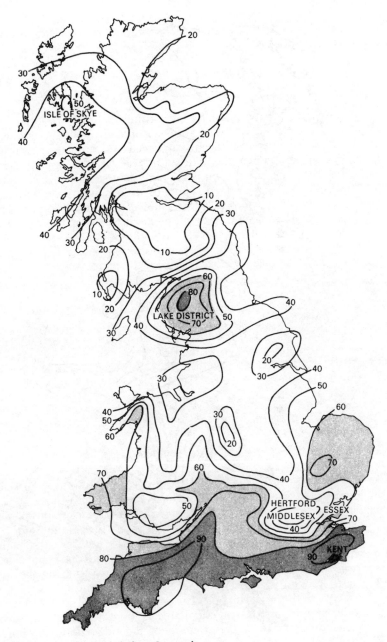

Figure 2.2 The view of girls from Sevenoaks.

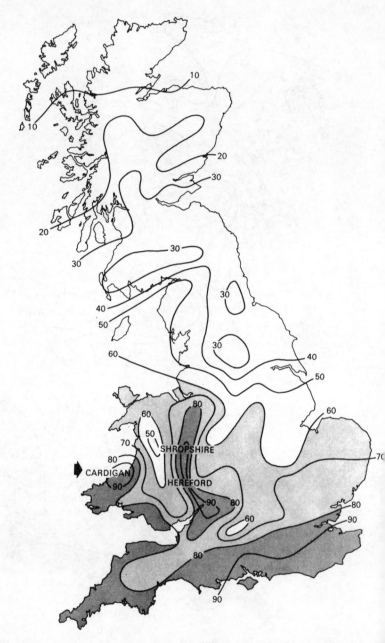

Figure 2.3 The view from Aberystwyth.

Marchlands of Wales, but extends south and east to include Somerset and Oxfordshire. Another remarkable difference, especially when we recall the view from Bristol, is the complete lack of a sharp gradient on the perception surface to the south-west of England. Certainly the dislike of the Bristolians is not reciprocated, as many Welsh school leavers perceive Somerset (Avon and Somerset) very positively. To the north the surface is much lower, and we must note that this was the only school out of the 23 sampled all over Britain which failed to show the striking Lake District Dome. It is almost as though the Welsh pupils are more than satisfied with their own rugged and beautiful landscape, and because they are satiated with their own mountains they view Westmorland (part of Cumbria) in a 'take-it-or-leave-it' fashion.

THE LIVERPOOL VIEWPOINT

Further north at Liverpool (Fig. 2.4), more familiar features of mental topography appear once again. Preference for the local area is very strong, and the peak over the perception point has a much gentler gradient to the north as it reaches out to coalesce with the Lake District Dome. To the south, we can recognise the usual ridge of desirability along the south coast. It is somewhat lower than we have seen so far over the south-east, but it rises in its now familiar way to a high point over Devon. The low-lying Mental Cirque over the Midlands is easily recognisable, and it is bounded in the west by the Border Prong which joins the local peak to the highly perceived south coast. With their western location, the school leavers from Liverpool appear to accentuate the western side of Britain at the expense of the east.

THE REDCAR VIEWPOINT

A similar, though opposite, bias is seen on the east coast, as we move north to Redcar in the North Riding of Yorkshire (Cleveland) (Fig. 2.5). Here the local peak around the perception point reaches south along the east coast to coalesce with the Anglian Prong and the Southern Ridge. The latter is perceived as extremely desirable, with values consistently over 90. Only the North Riding itself (Cleveland and North Yorkshire) receives a similar score, and the strong local effect reaches west to join with the Lake District Dome. To the north the gradient is very steep over the English–Scottish border, and we have, once again, striking evidence that the English in Britain tend to perceive other national areas as quite distinct, if not to say alien, countries. Notice, too, how the bias along the east coast is at the expense of the west (just the opposite of Liverpool), and how the

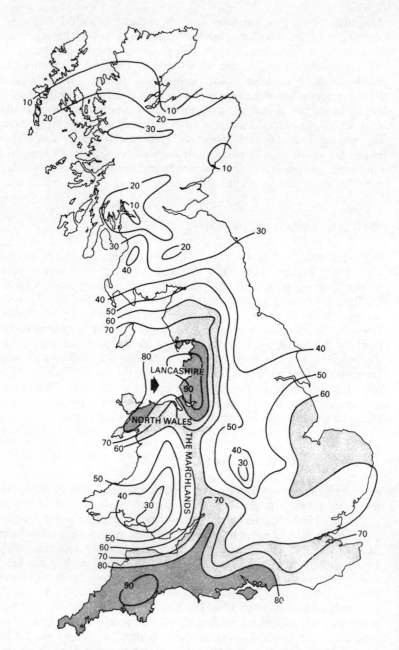

Figure 2.4 The view from Liverpool.

Figure 2.5 The view from Redcar.

Midland Cirque is now open to the unrelieved Welsh Depression. Once again, this is bounded by the extraordinarily sharp gradient between England and Wales over the Bristol Channel. Even though Redcar is nearly 400 km away, the English–Welsh gradient is seen as even steeper than the English–Scottish one much closer to home.

THE INVERNESS VIEWPOINT

Finally, the viewpoint shifts over the border into Scotland at Inverness (Fig. 2.6). Here the local effect appears extremely prominent as the high peak of local desirability reaches out in a prong to the east and Aberdeen. To the north the gradient is very sharp, for the inaccessible fastness of Sutherland appears as unattractive to Inverness eyes as to Sassenachs, but to the south the mental surface is quite convoluted. While all the other British viewpoints blanket the area with low scores, the Scots themselves discriminate quite carefully between the Scottish counties. We shall see this local tendency to discriminate carefully, in areas homogenised by others, once again when we look at some of the mental maps of students in the USA. Here the county of Midlothian (Lothian), containing Edinburgh, is a secondary Scottish peak which coalesces along a ridge of desirability with a third peak in Ayrshire (Strathclyde). Along the border with England, especially in the east, the gradient is quite sharp, and the North of England appears as unattractive to Scots as Scotland does to school leavers in Yorkshire. However, the condemnation of England is not a general one. To the south the familiar features of mental topography appear once again, with the south coast and East Anglia seen as moderately desirable. London does not appear too attractive to young Scottish men and women, and the view of Wales is so disparaging that it forms the low point on the entire surface.

Although each of these six viewpoints have their own distinctive characteristics, what is even more striking is the similarity between them. Many of the major features on the mental maps appear again and again, even though the perception points were at opposite ends of the country. It is almost as though we had some overall map of the basic mental topography moulded out of rubber, with a tennis ball underneath it. We could move the tennis ball around under our pliable map to the various perception points so that the lump pushed up represented a particular dome of local desirability. Perhaps the tennis ball would not have much distorting effect in the South of England, because it would be hidden under the generally high area of the south. But as we moved it northwards it might disclose the local effect more and more prominently.

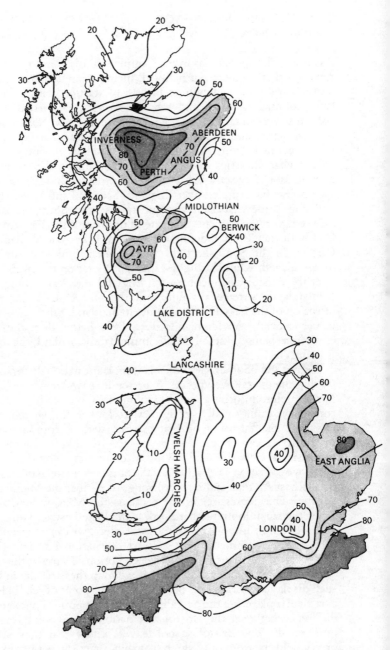

Figure 2.6 The view from Inverness.

THE NATIONAL SURFACE AND THE LOCAL DOMES OF PREFERENCE

Assuming the spatial bias in our sample schools is not too severe, we can construct a general, or national, mental map by combining all the individual school viewpoints. Instead of starting with a data matrix with counties as rows and school leavers at a particular place as columns, we can take each of the 23 school maps (we examined only 6 of them here), and treat them as columns in our matrix. The mental map we make from this information displays the most common, shared characteristics of all the individual maps that went into its construction. In a sense it is a weighted cartographic average of all the viewpoints we sampled over England, Scotland and Wales (Fig. 2.7).

Using somewhat Bunyanesque terms, the Southern Plateau of Desire is very prominent, and extending from it the East Anglian and Welsh Border Prongs define the now well known Midland Mental Cirque. In the South-East the Metropolitan Sinkhole can be easily distinguished, while the highest gradient by far on the entire surface is over the Bristol Channel between England and Wales. Once past the Midlands, the surface falls quite smoothly and regularly except for the intrusion of the Lake District Dome. In Scotland, where generally low values prevail, the Highlands do a little better than the rest, the exception being a small knoll immediately around the capital at Edinburgh.

We can get another view of this important national surface, representing the overall viewpoint, by pretending we are in a geographical Earth satellite in orbit about 300 kilometres over Oslo, Norway. Of course, we cannot really get up there to look at this shared, but quite invisible, mental view of Britain, but with a computer-mapping program we can choose any perspective we like to obtain a three-dimensional view (Fig. 2.8). With the computer helping us to get into orbit, we can see very clearly how the Southern Plateau of Desire sends out its East Anglian and Border Prongs to enclose the Midland Mental Cirque. The depression in the south is the Metropolitan Sinkhole, while the small lump further north is the Lake District Dome.

To reveal the importance of the local effect, we can undertake a bit of cartographic arithmetic, and subtract the national surface (Fig. 2.7) from each of the maps showing the perception surfaces at the six places we have examined (Figs 2.1–2.6). If we map these results in turn, we shall display the differences between the viewpoint of school leavers at a particular place and the overall, national viewpoint. For example, the difference between the Bristol and national viewpoint (Fig. 2.9) is quite small. The Bristol school leavers know that they are sitting pretty, and they can only agree that they share the national view to a considerable extent. Small plus and minus values are scattered over the

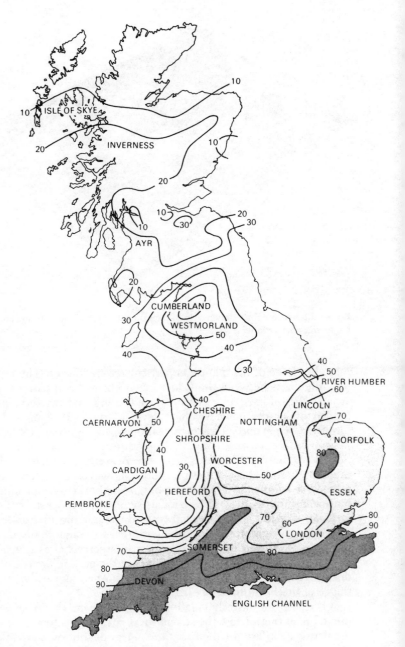

Figure 2.7 The general, or national, perception surface.

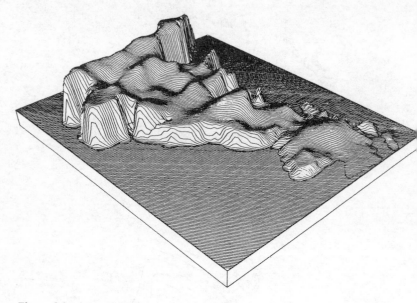

Figure 2.8 A computer-drawn map of the national surface, viewed from about 300 kilometres over Oslo, Norway.

map, but none of them appears particularly significant. The residuals (or differences between the values on the Bristol and national maps) around Bristol itself are very small, indicating that any local effect is hidden under the generally high level of the Bristol mental map. Similarly, when one subtracts the national perception surface from the Sevenoaks mental map (Fig. 2.10), the residual values of Kent and Sussex are almost zero. Once again, the local effect is hidden under the high Southern Plateau of Desire. Notice, however, that there are three quite large anomalies. The Metropolitan Sinkhole is very prominently outlined with high negative residuals, while Brecknockshire (Powys) in Wales, and the Isle of Skye in Scotland, are both positive anomalies. The Welsh anomaly may be nothing more than a random discrepancy, but in Scotland the whole highland area is perceived as much more desirable than the national viewpoint, and the peak in the Isle of Skye does suggest that the girls of Sevenoaks have an unusually romantic image of this beautiful area of Britain.

When we subtract the National Surface from the Aberystwyth mental map (Fig. 2.11), the strong local peak begins to appear above the national surface for the first time. Most of southwestern Wales is perceived as well above the national view, while a prong extends through Shropshire to Flint (part of Clwyd). Warwick is a small

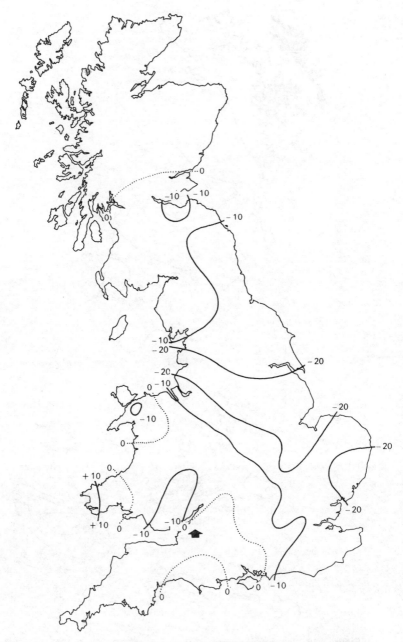

Figure 2.9 The difference between the national and the Bristol perception surfaces.

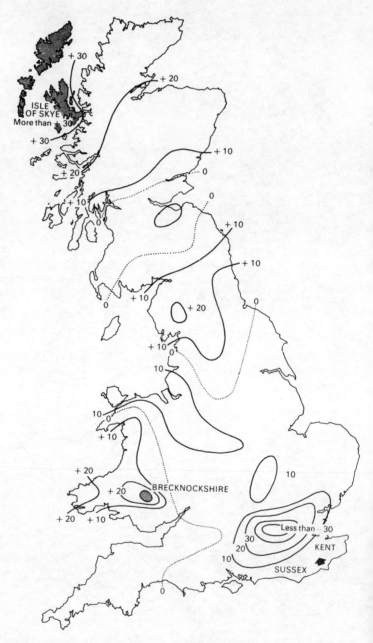

Figure 2.10 The difference between the national and Sevenoaks perception surfaces.

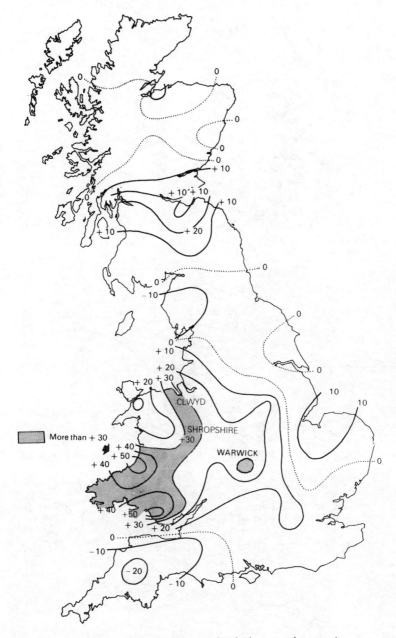

Figure 2.11 The difference between the national and Aberystwyth perception surfaces.

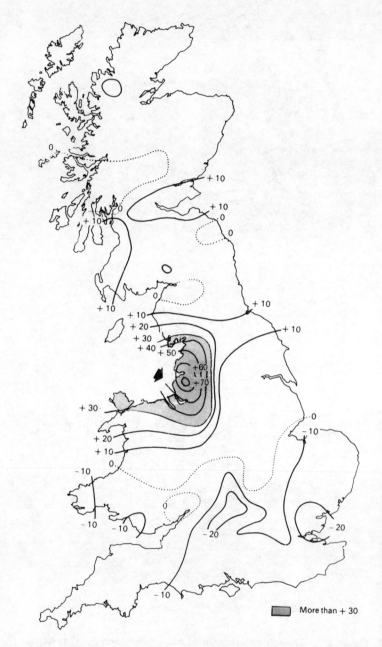

Figure 2.12 The difference between the national and Liverpool perception surfaces.

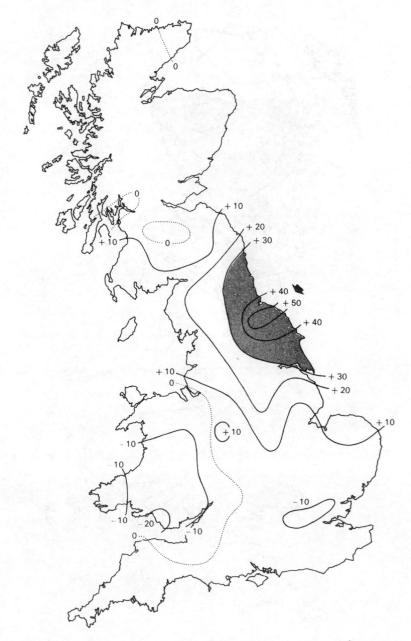

Figure 2.13 The difference between the national and Redcar perception surfaces.

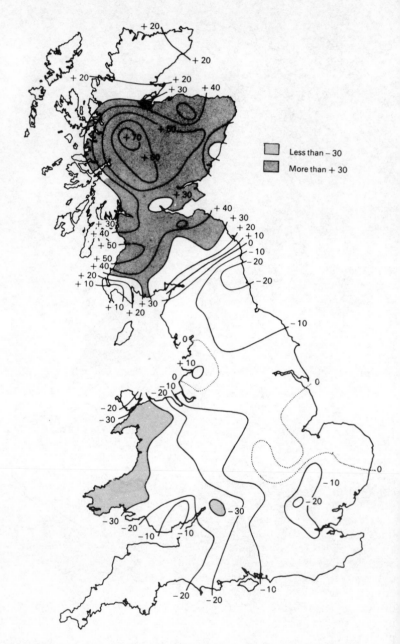

Figure 2.14 The difference between the national and Inverness perception surfaces.

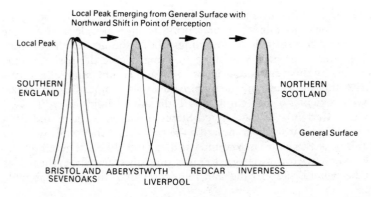

Figure 2.15 A cross-sectional view showing the emergence of the local dome with the northwards migration of the perception point on the general surface.

residual outlier, but the local effect is so marked that there is a consistent distance decay effect with movement away from Aberystwyth. The same is true when the perception point moves north to Liverpool (Fig. 2.12). The local dome has pushed well above the national surface, and the trend is consistently down as one moves in any direction away from the city. Everywhere else the residual values are very small, indicating that it is virtually only the local effect that is responsible for any difference between the national and Liverpudlian viewpoint. At Redcar (Fig. 2.13) the local dome again pokes up prominently above the national surface, and by the time Inverness is reached it is almost completely exposed (Fig. 2.14). The dislike of the Scottish school leavers for Wales is also prominently displayed as a strong negative anomaly. Even though Wales is not a high area on the national surface, it still appears as an area of marked negative residuals, a region viewed in a very disparaging light indeed.

The growing prominence or exposure of the local domes of desirability with a northwards shift is so regular and marked that we can summarise everything we have seen with a very simple graphic model. If we think of taking a slice through Britain in a north–south direction (Fig. 2.15), we can represent the National Perception Surface as simply a wedge that is high in the south of England but low in the north of Scotland. Our local domes of desirability are sharp peaks centred on the points from which the views are being taken. When the perception point is in the south, say Bristol or Kent, the local effect is not seen as a distinctive feature above the national surface. But as the point moves northwards the local dome of desirability becomes progressively more prominent, until it is almost completely exposed at Inverness in Scotland.

Thus, any mental map of school leavers from a particular place is a rather subtle convolution of (1) a shared, national viewpoint and (2) a dome of local desirability, representing feelings people have for the familiar and comfortable surroundings of their home area. If you are a British reader, perhaps you share to some degree the basic feelings expressed by this very simple abstraction and compression of reality – in brief, this model of a mental map. And if you are not a British reader, to what extent do you think this little model holds in your own country? Certainly in the United States, Nigeria, Sweden and Peninsula Malaysia it seems to describe the situation rather well. We shall see this now as we turn to a country with a quite different geographical scale, where we can also begin to raise some questions about the things that lie behind these strong and regular mental images of habitable space.

3 Environmental preferences and regional images in the USA and Canada

'South Dakota!' Inigo's cry was ecstatic. The man must really have been there because you couldn't *think* of South Dakota, couldn't just lift it out of some mental map.

J. B. Priestley, *The good companions*

When we leave Britain and turn to the USA to examine the varying mental images that people hold, we must remember that our geographical scale increases by at least an order of magnitude. While we would not like to encourage the ingrained propensity of Texans to exalt sheer size, we should realise that the whole United Kingdom could almost fit three times over into that state alone. The overall picture we generate of people's preferences, and some of the things that may lie behind them, will necessarily be painted with a broad brush in rather large strokes. The USA is so large that it is simply impossible to obtain preferential information from people with the same 'fine-grained', spatial resolution of Britain. We would literally require people to rank thousands of sub-areas, and the task is conceptually impossible and much too time-consuming. Although the scale of our inquiry has greatly expanded, we might suspect from our previous examination of geographical images in Britain that the American mental maps will also share some general features, while displaying some particular individual properties depending upon where the perception points are located.

We are going to look at the mental maps of university students at five locations: four in the North and West of the USA, and one in the Deep South (Gould 1965). The geographical images and preferences of university students may be extremely important in the future, for we must realise that these young men and women are the most highly educated and mobile members of an incredibly mobile society. One of the things that strikes many Europeans (not to say many Americans themselves) is the extraordinary willingness of individuals and young families to pick up sticks and move literally thousands of miles to new places and different homes, surrounded by new neighbours and unfamiliar faces. Often these moves are made at quite short intervals of a year or two, as new opportunities for well educated people open up frequently, or as firms transfer their young management personnel to various branch plants and marketing areas around the country to gain experience. People in government and the academic world also tend to be much more mobile than their peers overseas; often with detrimental results, for these occupations require a nice balance between the stability of settled and involved people and the numbing stagnation of tenured sinecures. We might also speculate that one of the reasons for the general deterioration and distressing visual blight of many American towns is simply that the young and well educated are so mobile that it is difficult for many of them to make an involved commitment to local politics. The result is that entrenched commercial interests tend to be well represented locally, and so produce the grabbingly visual urbanscapes and billboard alleys that characterise so many towns and their once-beautiful approaches.

THE CALIFORNIA VIEWPOINT

The first group of students we shall consider are located in a state that is generally one of the most highly regarded in the country: California, with students at Berkeley near San Francisco. Working with a map of the 48 contiguous states, and ranking them in the now familiar way, this group has no doubt about the desirability of California itself (Fig. 3.1). The peak of the preference surface lies over the state, and a high ridge of desirability runs along the entire west coast into Oregon and Washington. The gradient to the east, however, is very sharp, as preferences decline rapidly into a mental basin centred over the states of Nevada and Utah. The latter's image is quite tarnished, perhaps because most people do not like the degree to which one religious group, the Mormons, dominates virtually every aspect of life there. It is no fun residing in a state in which you will always be considered an outsider. Moving further east, the surface rises once again to a secondary peak over Colorado, and then drops steadily to the states of the Great Plains. These generally have a rather poor image when seen from California, and the Dakota Sinkhole forms one of the lowest areas on the entire surface.

A particularly interesting aspect of this mental map is the way the entire surface changes its general orientation as it moves eastward through the Great Plains states. Whereas the general trend towards the east was downwards from California, now the surface rotates 90°, and the overall trend is in a north–south direction, with a Southern Trough forming the lowest point on the mental map. Alabama, Mississippi, Georgia and South Carolina are the last places in the country for California's students. Only Florida escapes the general condemnation of the South, although Californians do not perceive this state as highly

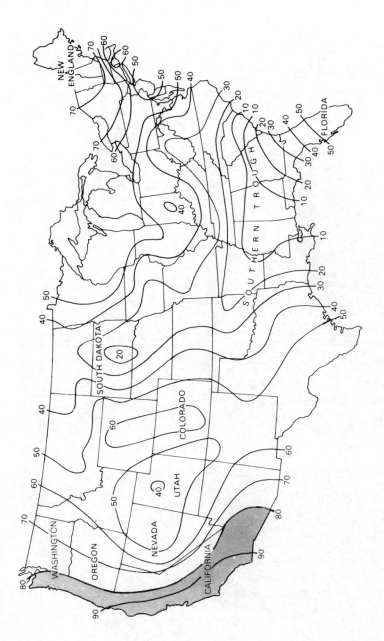

Figure 3.1 The mental map from California.

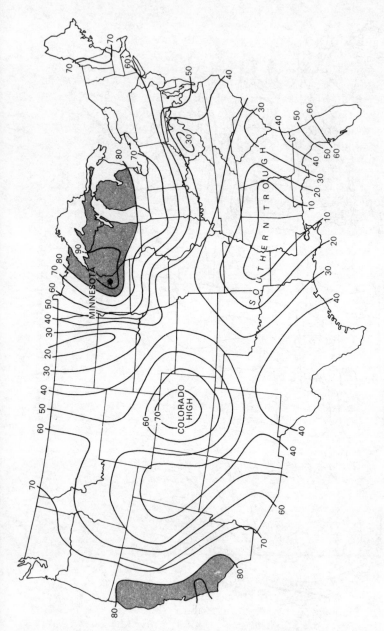

Figure 3.2 The mental map from Minnesota.

as some other groups of students in the North – perhaps because of an old rivalry for the title of America's premier place in the sun. Northwards, the surface trends consistently upwards to the states of the Middle West and New England, although it is worth noticing the way in which West Virginia distorts the even march of the isopercepts. Problems of poverty, rapacious strip-mining, and an appalling lack of decent medical care in this area of Appalachia, are all well publicised aspects that form part of the information available to this group of socially aware students. Much of Kentucky is in a similar social and economic plight, but her perception score is nearly twice that of her northeastern neighbour. Perhaps white-fenced, bluegrass pastures, and sour mashed bourbon with pretentions to spirituous greatness, familiar themes in many advertisements, have moulded a mental image brighter than the facts over much of the state would warrant. From Pennsylvania, north-east to New England, the rise is steadily upwards to a third peak that seems to carry visions of the mountains and lakes of Vermont and New Hampshire, and the quiet, rock-strewn coast of Maine.

THE MINNESOTA VIEWPOINT

If we now move the perception point about 2200 km to the east, to sample the mental maps of university students at the University of Minnesota (Fig. 3.2), the overall mental map is extraordinarily similar, despite this massive shift in location. A local dome of desirability forms the primary peak of the surface over Minnesota, but there is also a high ridge of desirability along the west coast reaching out from a secondary peak in California. Once again, the Utah Basin and Colorado High are quite distinct, and the steepest gradient on the entire surface is west to the Dakota Sinkhole. The Southern Trough is the lowest area, and the surface rises slowly to the Mason–Dixon line, the traditional division between the North and the South, and then more rapidly to the States of the Midwest and the East. Even the local West Virginian Depression can be recognised once again. Except for the local domes of desirability over the perception points, the mental maps of Californians and Minnesotans appear virtually identical.

THE PENNSYLVANIA VIEWPOINT

The same map appears when we shift about another 1600 km east to Pennsylvania (Fig. 3.3). A high local dome centred in Pennsylvania also covers New York, Connecticut and Massachusetts, and elsewhere the surface is virtually identical to the Californian and Minnesotan

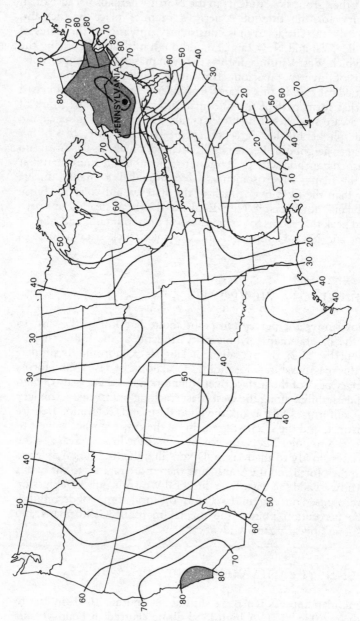

Figure 3.3 The mental map from Pennsylvania.

examples. The Western Ridge falls and rises through the Utah basin and Colorado High; the Dakota Sinkhole is just as prominent; and the nadir of residential desirability is over Alabama and Mississippi in the Southern Trough, with only Florida escaping the general southern evaluation. Notice, once again, how the steepest gradient in the East lies over the Mason–Dixon line, particularly south-west from the Pennsylvanian Peak to the West Virginian Depression.

Despite locations thousands of miles apart, this generation of Americans appears almost unanimous in its perception of geographical space. The three examples we have seen so far, from the North, East and West of the United States, tend to confirm the ideas we developed from the mental maps in Britain. There may be a general or national surface representing the shared viewpoint of most people in the country, together with the local dome displaying strong preferences for the immediate area. Our analogy of the national surface moulded from rubber, with a tennis ball pushing up a local dome of preference, seems to hold at this greatly enlarged geographical scale.

THE ALABAMA VIEWPOINT

Yet if we know anything at all about the rich diversity of people, cultures and landscapes in the USA, there is something almost too good to be true about such a pat simplification. And so it turns out to be when we move the perception point south to sample the mental maps of white university students at the University of Alabama (Fig. 3.4). Here we have a very different view, for parts of the map are in marked contrast to those we have just examined in the North. As we might expect, the highest point is the local dome of desirability centred over Alabama, but the truly surprising thing is the very steep gradient west to the neighbouring state of Mississippi. Students in the North do not discriminate between these adjacent states, and mentally homogenise them into the Southern Trough. In contrast, Alabamans discriminate very carefully between states in the South, and rate Louisiana and North Carolina well above such neighbours as Mississippi and South Carolina. Clearly the Alabamans have much more information, and undoubtedly more travel experience in the South, and make careful comparative judgements in an area that others blanket with low perception scores. This is very similar to the way in which the Inverness students discriminated carefully in Scotland, an area consistently perceived as a low plain of dislike by students in England and Wales.

In the North the surface is also quite different from anything we have seen before. The trend is consistently down to 'damned Yankeeland', and clearly no self-respecting southerner would be seen dead in such a

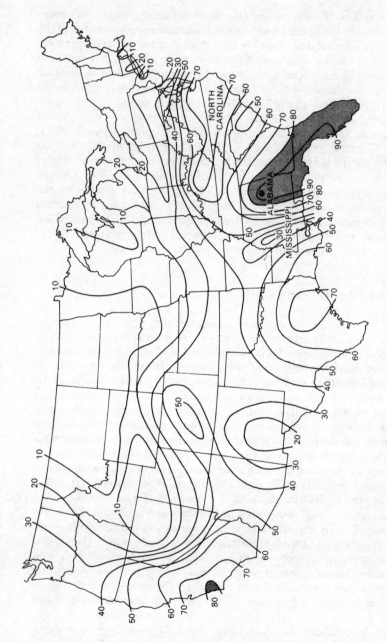

Figure 3.4 The mental map from Alabama.

perfidious and alien area as New England. The western states, however, did not take part in the Civil War over a century ago, and here the perception surface conforms once again to the now familiar pattern. There is a moderate secondary peak over California, and features such as the Utah Basin and the Colorado High are easily recognisable. The Dakota Sinkhole is now only part of a vast area of northern dislike, for the southerners also have a propensity for mentally homogenising areas far from their home territory.

THE NORTH DAKOTA VIEWPOINT

For our last example we are going to shift northwards once again and examine the mental map of students in North Dakota (Fig. 3.5). While it conforms in many ways to the general northern type, it is the only example we have found so far that does not possess a clear local dome of desirability. Indeed, the Californian and Colorado peaks are distinctly higher, and even Minnesota and Wisconsin are scored as more preferable. Little wonder that these Great Plains states are areas of out-migration, with the young and well educated people leading the way. Notice, also, the great contrast between the mental maps of North Dakotan and Minnesotan students (Fig. 3.2). The views they have of each other are hardly reciprocated, and the contrast reminds us strongly of the Bristol and Aberystwyth pair in Britain, where the steep perceptual gradient from Bristol northwards across the channel to Wales did not appear at all on the Aberystwyth map looking the other way.

Apart from the local-dome anomaly, the mental map from North Dakota conforms well elsewhere to the generalisations we made earlier. The Utah Basin, between the Californian Ridge and Colorado High, is quite distinct, and there is a clear downwards trend to the Southern Trough before the surface rises to moderately high values over Florida. The eastern states also share these moderate values, although New Jersey is somewhat lower, perhaps because the sample was taken shortly after large-scale riots in Newark in 1968 and the rank assessments were affected by these well publicised civil disturbances.

At this much larger geographical scale our simple graphic model of a shared national perception surface, convoluted with a local dome of desirability, stands up reasonably well for the northern and western viewpoints. It is obvious, however, in an area of such size, diverse culture and historical experience, that sometimes our simplification is going to break down. The southern mental map is totally different – indeed, the complete inverse – of the northern view over the eastern part of the United States, and it is worth noting the way these deeply rooted regional images confirm, once again, the division of the country

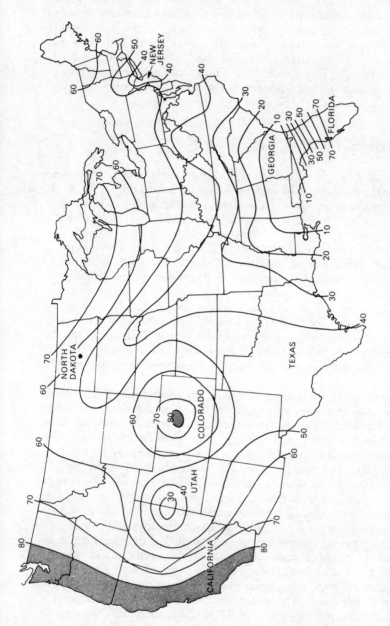

Figure 3.5 The mental map from North Dakota.

along the tragic lines drawn over a century ago. Only in the west do the southern views conform to the overall national surface – the one area that was not really deeply involved and committed in the Civil War. Our model of national and local effects is also a bit shaky when it comes to the North Dakotan view. Clearly some areas are so unfortunately placed and blessed, relative to other parts of the country, that even their own inhabitants take a rather sour view of them. In North Dakota the strong local effect is considerably weakened as geographical allegiances and high-ranked values are mentally assigned to other areas.

What lies behind these mental images of geographical space – images that for the most part are extraordinarily stable and predictable? This is a very difficult question to answer at the moment in any really satisfactory way, and even now a number of geographers, and their colleagues in other disciplines, are continuing their research into these puzzling areas of of geographical perception. When we looked at the construction of perception surfaces in Chapter 1, we showed in a rather schematic fashion the flows of information that people receive and order into lists of residential preferences (pp. 28–9). We shall examine questions of information and ignorance at some length a little later in Chapter 4, for these are questions concerning the *input* to the 'black boxes' of Figure 1.14. But we can also gain some insight into these questions by examining other forms of geographical output, as people respond to questions of evaluating the states in America.

PERCEPTION SPACE AROUND TEXAS AND GEORGIA

One way of probing behind these rather consistent images of geo-graphical space is to let people respond in a very open and unstructured way to the states themselves. Two geographers, Lloyd and Van Steenkiste, ran such an experiment by presenting the names of each of the 48 contiguous states in a random order to large groups of university students in Texas and Georgia. As each name appeared, the students were asked to write down whatever words came into their heads. The words were then coded in a rather loose classification to form 26 categories ranging from physical characteristics, through social atti-tudes, aesthetic comments, recreation and sports, and so on. We can imagine a matrix 48 × 26, in which each state in a row is characterised by scores on the 26 categories in the columns. By using a method which is a simple extension of the one we used earlier to obtain perception or residential desirability scores, it turns out that many of these categories collapse on to two quite independent axes or rulers. On one ruler, such things as landforms, climate, flora, and outdoor recreation and sports load very strongly; while on the other place names, social characteristics, cultural landscape, pollution and per-

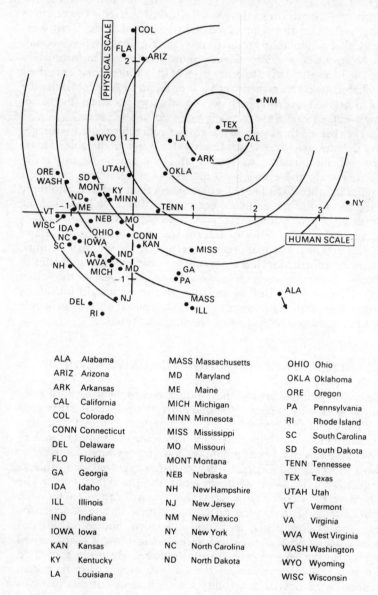

Figure 3.6 Texas, and other states of America, located in 'perception space'.

ALA Alabama
ARIZ Arizona
ARK Arkansas
CAL California
COL Colorado
CONN Connecticut
DEL Delaware
FLO Florida
GA Georgia
IDA Idaho
ILL Illinois
IND Indiana
IOWA Iowa
KAN Kansas
KY Kentucky
LA Louisiana

MASS Massachusetts
MD Maryland
ME Maine
MICH Michigan
MINN Minnesota
MISS Mississippi
MO Missouri
MONT Montana
NEB Nebraska
NH New Hampshire
NJ New Jersey
NM New Mexico
NY New York
NC North Carolina
ND North Dakota

OHIO Ohio
OKLA Oklahoma
ORE Oregon
PA Pennsylvania
RI Rhode Island
SC South Carolina
SD South Dakota
TENN Tennessee
TEX Texas
UTAH Utah
VT Vermont
VA Virginia
WVA West Virginia
WASH Washington
WYO Wyoming
WISC Wisconsin

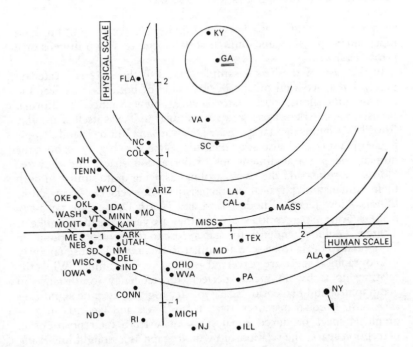

Figure 3.7 Georgia, and other states of America, located in 'perception space'.

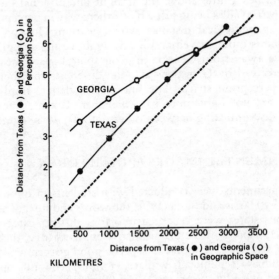

Figure 3.8 The way distances in geographical and perception spaces vary together for the Texas and Georgia samples.

sonalities contribute very heavily. Obviously, we have strong and independent physical and human scales forming a two-dimensional *perception space*.

In the case of the Texas sample (Fig. 3.6) the state is located in perception space well on the positive side of both the physical and human dimensions, with states such as New Mexico, California, Louisiana and Arkansas seen as very similar to Texas itself. Far away from Texas in this mental space are Delaware and Rhode Island. People in Georgia (Fig. 3.7) also scale their own state very highly on both the human and physical dimensions of perception space, and they see Florida, Virginia and the two Carolinas as being quite similar in both their human and physical characteristics. Conversely, New York, Illinois, New Jersey, Rhode Island and North Dakota are far away from Georgia in perception space, for they are viewed as most unlike it. There is a definite tendency for states close to the perception points in Texas and Georgia to be seen as similar to them, while those far away in geographical space are perceived as being very dissimilar. While the relationship is by no means perfect, the tendency for distance in geographical and perception space to vary together is quite significant (Fig. 3.8). If equal distances away from Texas and Georgia in geographical space produced equal steps away from their positions in perception space, the relationship would graph as a straight line rising at 45° from the origin of the graph. In fact, the Texas line is slightly curved, and lies a little above the straight line of equal increments. States are perceived as being just a bit further away in perception space than their geographical distance would warrant. In contrast, the Georgia line is quite curved and lies well above the straight line. A small move away from Georgia in geographical space means that a state is perceived by Georgians as quite different, and therefore far away, in perception space. The line is also flatter, implying that Georgians are not only more parochial than Texans, but also less capable of discriminating between other states far away from them.

ENVIRONMENTAL IMAGES FROM ILLINOIS

Similar experiments were conducted with students in Illinois. In the early tests by Harris and Scala (1971) the words and images associated with various states were classified into four distinct categories that described (1) aspects of the physical environments (dry, dusty, cold, mountains, flat, etc.), (2) political aspects (good government, corrupt cities, racial bias, Republican, Wallace, etc.), (3) economic opportunities (no jobs, wealthy, poor, etc.), and (4) social and cultural aspects (snobbish, hospitable, clannish, etc.). From this first classification, four scales were devised for another round of questions posed by

Harris *et al.* (1970) in a second experiment. Each scale had seven points on it. For example, a person who regarded Colorado very highly on the environmental scale might award this state the highest value of seven, while scaling New Jersey only one for a very unattractive physical setting. On the political scale another person might evaluate Illinois one, North Carolina a middling four, and Minnesota a high seven – and so on, for each state on each of the four basic scales.

Once again, we have a considerable amount of information about people's geographical preferences, but this time on four basic scales that perhaps will begin to tease out some of the main, underlying threads in the rather consistent spatial patterns we have seen before. For example, we can calculate a single score for each state on the environmental scale alone (Fig. 3.9), and think of these values as crude measures of the degree to which the students' geographical images vary when they consider this important aspect of an area. Notice how most of the country lies between the very high (over 80) and very low (under 20) enviro-preferential contour lines. People appear to perceive most of the states as not too bad and not too good as far as the physical environment is concerned. Perhaps a lack of information accounts for these fairly wide areas of environmental indifference; or perhaps many states are so large that they bring to mind a great variety of environmental images. In contrast, the extreme areas are quite sharply defined. California and Colorado (as seen from Illinois, remember) are highly regarded as very attractive physical settings. On the other hand, a broad belt of states of extraordinary physical contrast, ranging from the hot Deep South to the flat Dakotas with their summer and winter extremes, is condemned as physically almost uninhabitable. Another area of physical disparagement lies over Utah and Idaho, both of which have wide ranges of terrain and climate. Here dramatic desert landscapes merge into much more moist areas of majestic mountain scenery, yet the areas are lumped together and apparently condemned for their physical unattractiveness.

On the political scale (Fig. 3.10) the same wide areas of indifference occur, but this time more emphasis is placed upon the positive side of the scale. California, Colorado, the Upper Midwest and most of the North-East, are regarded as having favourable political climates, while only the Deep South is seen in a really unfavourable light. While the spatial pattern is certainly not identical to that of the physical environment, the striking correspondence of some of the highs and lows is worth noting. Similarly, on the economic scale (Fig. 3.11), the spatial pattern of perceived economic opportunity closely resembles the environmental and political maps. California and the North-East are regarded favourably (Colorado is a smaller peak), and the South and Great Plains are viewed very disparagingly. Finally, on the social and cultural scale (Fig. 3.12), the now familiar overall pattern appears once

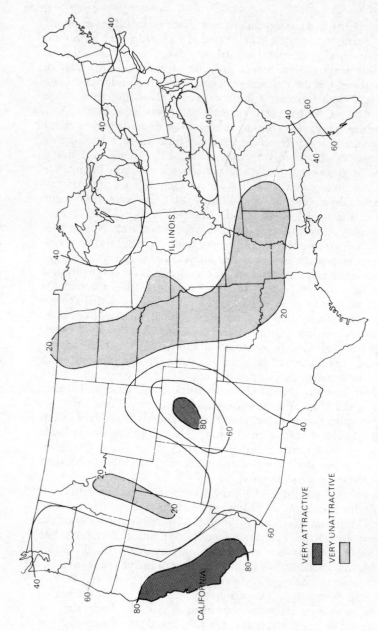

Figure 3.9   The mental map of Illinois measured only upon the environmental quality scale.

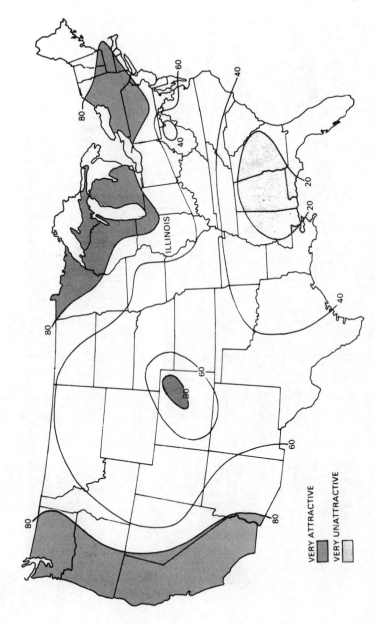

Figure 3.10 Illinois: the mental map measured on the political quality scale.

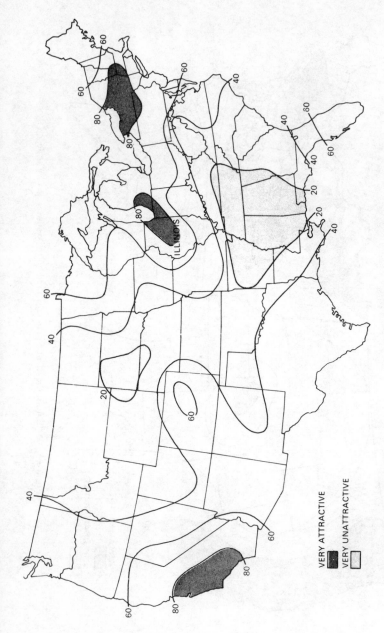

VERY ATTRACTIVE

VERY UNATTRACTIVE

Figure 3.11 Illinois: the mental map measured on the economic opportunity scale.

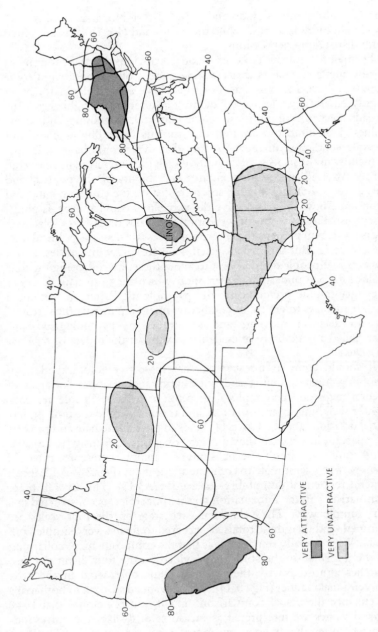

ILLINOIS

VERY ATTRACTIVE

VERY UNATTRACTIVE

Figure 3.12 Illinois: the mental map measured on the social scale.

again. California, Colorado and the Midwest–North-East all receive moderate to high scores, while the South and Great Plains are pitted with disparaging depressions.

From our simple visual comparison it is obvious that these four maps hardly measure aspects that are independent of one another. People appear to have a strong tendency to evaluate a particular state, either highly on all the scales, or condemn it as undesirable right across the broad spectrum of political, environmental, economic and social values. It is almost as though the students from Illinois could not tolerate conflicting images – images that they would have to handle somehow in the process of defining which they preferred in an overall sense. After all, if you view an area very highly on the political and social scales, but feel it is far down the list on the environmental and economic, how do you resolve the perceptual conflict? At this point, you must start a process of weighting the scales themselves, placing a higher value on perhaps the environmental scale over the political one, or vice versa. But this process itself generates great difficulties, and it appears that people do their best to avoid these, by collapsing the four scales into one underlying ruler upon which they can happily order their geographic preferences. The psychologist Patterson has noted such tendencies to get rid of conflicting images in many other areas of human decision making, and a number of psychologists have marshalled a large body of evidence to support their ideas of *cognitive dissonance*.

It should come as no surprise that these psychological ideas of resolving mental conflict and uncertainty should appear in areas of mental maps and geographical perception surfaces. People are constantly weighing up information, and trying to come to some preferential decisions about all sorts of things in their immediate and broader life spaces. However, if they do carry over this propensity to collapse conflicting scales into these areas of geographical preference, it would seem perfectly legitimate to combine all the maps (Figs 3.9–3.12) into a general residential desirability surface (Fig. 3.13). This overall map, summarising all the information on the four scales, is constructed in a very simple way. The values for each state on the four scales are summed, and then these totals are ordered to form a very simple rank score. What is really remarkable is that the composite map, collapsing all the information the students provided on the four distinct scales, matches almost exactly their mental map constructed in the now conventional manner (Fig. 3.14). We have added shading to both maps to enhance the visual comparison, and if you remember that large ranked values are interpreted as measures of dislike, the correspondence between the two perception surfaces is most striking. The West Coast Ridge falls through the Utah Basin before rising again to the Colorado High, and the perceptual valley linking the Dakota Sinkhole to the Southern Trough is very marked on both. The peaks of the surfaces lie over Illinois itself, while the whole of the Upper Midwest and North-East are also seen as most desirable. It is precisely what we would predict from our knowledge of the national surface and the local domes of preferences, when the perception point is in the northern part of the United States. Both these roads to measuring the mental map – one via four collapsing scales to avoid cognitive dissonance, the other via the conventional ranking of preferences – lead to the same stable and highly predictable geographical image.

Although we have greatly altered our geographical scale in moving from Britain to the USA, our rather simple model of a national surface and a local dome seems to hold up fairly well. At the same time, we have started to look at some of the things that might lie behind these consistent and predictable images. In the USA, at any rate, people seem to evaluate an area on four scales, but collapse these into one to avoid the difficulty of dealing with conflicting images and information. We also have evidence that distances in perception space and physical space are positively related. Thus, most people tend to see the most distant states as being least like their own home state.

In a number of cases we have also seen how various groups of people discriminate in different ways within an area. Both the Alabaman and Inverness groups, for example, distinguished very carefully in the South and Scotland respectively – both areas mentally homogenised by people outside these regions. Now, in order to discriminate carefully between things we must have *information*. Indeed, we might suspect that information itself varies in rather consistent ways over a country. Clearly the Alabaman students had travelled much more in the South, and had acquired much more information about it than their northern peers. They had also read southern newspapers, which naturally recorded much more southern news than the usual papers in the North. Similarly, the Southern press and regional television provides much more information to the young men and women in Inverness about Scotland than the BBC and English newspapers give to their listeners and readers. In brief, there may be a distinct spatial bias to the information we acquire about our local areas, our countries, and the world. It is time we looked at these information inputs a little more closely and systematically.

ONTARIO AND QUEBEC

Important differences in attitudes to people and places persist in the USA. It is divided by race, and many residents speak Spanish as their only tongue. In Canada there are two official languages, a provincial party was elected in 1976 on a separatist platform, and secession was a

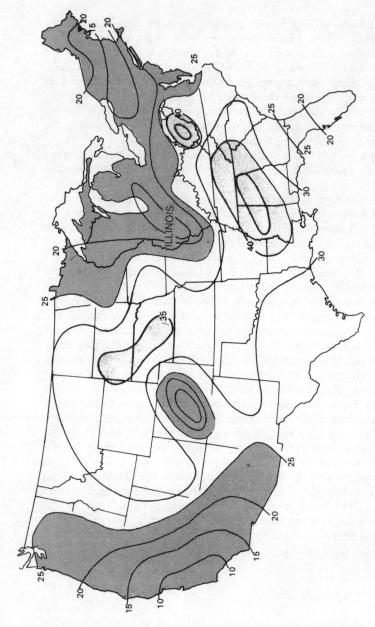

Figure 3.13 The composite mental map from Illinois combining the scores on the environmental, political, economic and social scales.

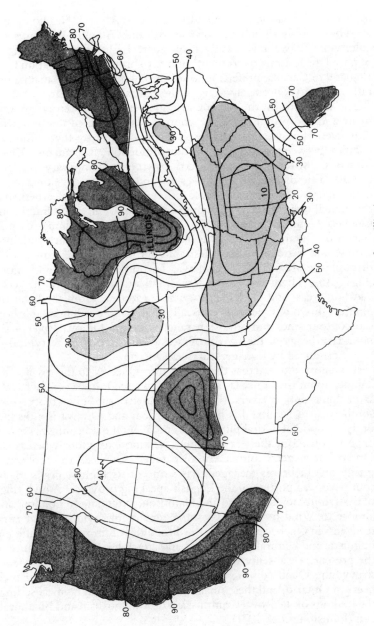

Figure 3.14 The standard mental map from Illinois.

topic of conversation from Newfoundland to Alberta. Headlines such as 'The West is sliding out of Canada' and 'Atlantic Provinces – a colony of Ottawa' raise hardly a comment. Eight of the ten provinces opposed the central government's move to bring the constitution from England to Canada, preferring to see the colonial anomaly maintained rather than to concede anything to the central government.

The key factor in the current struggle between the central government and the provinces is the control over resource development. However, this struggle is eclipsed in importance by the continuing conflict between francophone and anglophone Canadians. Deep-rooted in history, and reinforced by the contemporary resource dispute, this distinction continues to undercut the development of national unity. Nowhere are these differences seen more clearly than in the mental maps of teenagers in the adjacent provinces of Ontario and Quebec (Gould & Lafond 1979). In the small anglophone town of Bancroft, Ontario, 16-year-olds blanket their own province with high scores, and a ridge of high desirability runs from their own town through the urban areas of Oshawa, Toronto, Waterloo and London (Fig. 3.15). There are also a few small outliers, mainly attractive vacation areas, and the federal capital at Ottawa. From the latter the gradient down to Quebec is especially sharp, and it is remarkable how the contour lines run almost parallel with the Ottawa River – the boundary between the two provinces. This map is quite typical of many produced by teenagers all over Ontario.

If we now move across the border into Quebec, to the town of La Tuque, about the same size as Bancroft, we tend to see somewhat the same thing, only in reverse (Fig. 3.16). Much of the St Lawrence valley south-west to Montreal is highly regarded, and Ottawa is a distinct outlying node of desirability. So far, the mental map conforms to our expectations. But there is one quite striking exception: whereas the Ontario teenagers completely disparaged Quebec, this collective image of Quebecois teenagers has some of its highest peaks in the south-western portion of the English-speaking province. A high ridge of desirability joins Toronto and Hamilton, extending to London, and almost to Windsor. We have some remarkable evidence that rural teenagers in Quebec play down the difficulties of fitting into a different linguistic and cultural community, and attach far more importance to the presence or absence of urban areas. We have additional evidence that young Quebecois are more adaptable than their parents or their peers in Ontario, and these images confirm, in a geographical setting, the findings of the Royal Commission on Bilingualism and Biculturalism (Johnstone et al. 1971).

We cannot help adding a note of caution, however, since our remarks may only apply to young people in *rural* areas. These who grow up in big cities may form different, and in a sense harder, views,

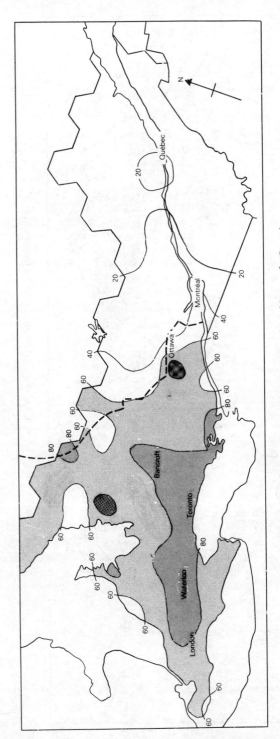

Figure 3.15 The mental map of 16-year-olds of Bancroft, Ontario.

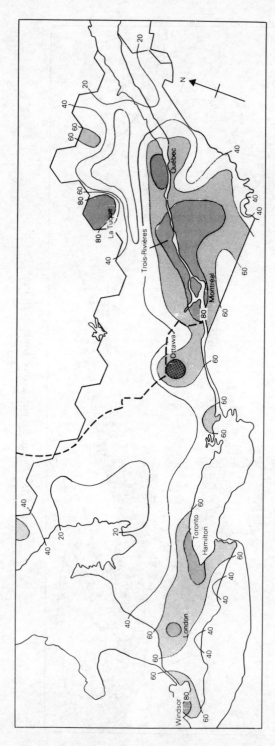

Figure 3.16 The mental map of 15-year-olds at La Tuque, Quebec.

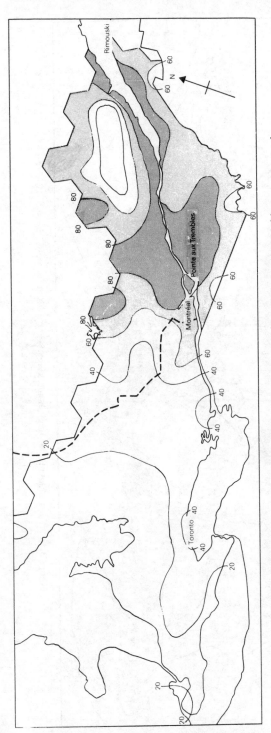

Figure 3.17 The mental map of French-speaking 17-year-olds at Pointe-aux-Trembles near Montreal.

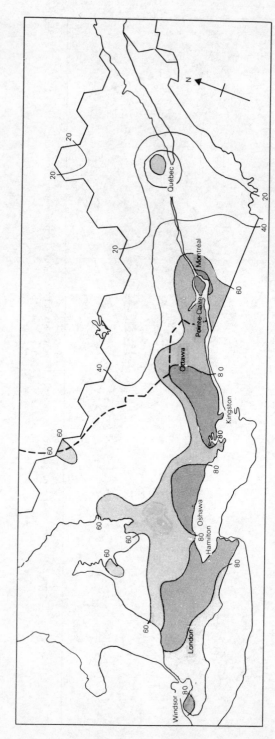

Figure 3.18 The mental map of English-speaking 15-year-olds at Pointe-Claire near Montreal.

s we see from two remarkable mental maps from Montreal (Gould & Lafond 1979). The first is that of 17-year-olds at Pointe-aux-Trembles, a French-speaking school at the eastern end of the island (Fig. 3.17). There is a strong bias towards Quebec, with the exception of very modest outliers at Toronto and Kingston. In marked contrast, we also have the mental map of teenagers at the English-speaking school of Pointe-Claire at the western end of the island (Fig. 3.18). Although they live in Quebec, and in the largest French-speaking city of Canada, these young anglophones have a quite decided preference for Ontario – with the exception of familiar Montreal, and a small outlier at Quebec City. Thus, we see that we have to be very careful about making generalisations in this complex area of mental images. These young people are only a few kilometres apart in geographical space, but literally leagues apart in mental space. As we shall see later (Chapter 4), these spatial and cultural biases appear to be strongly shaped by the very different streams of information to which these two groups are exposed.

4 Patterns of ignorance, information and learning

'Isn't this ridiculous – I don't even know what state we're in!'

Unsolicited remark from an
anonymous student in Illinois

When we examine patterns of spatial information and locational ignorance, we must immediately acknowledge one rather uncomfortable and discouraging fact: information is a very difficult thing to measure. We all tend to throw the term around in rather glib ways, yet when we try to pin it down, and subject it to closer analysis, we find it has very elusive properties. Like a number of other 'obvious' concepts in geographical enquiry (recall our discussion of accessibility in Ch. 1), we may have to be content with rather crude, and perhaps even naïve, surrogate measures.

We must also add a second caution here, for when you examine some of the appalling expressions of locational ignorance later in this chapter, you may well feel that much of what we have said about geographical preferences is jeopardised. In fact, we will argue that there need be no direct correlation between preferences and accuracy of location, nor between imageability and the task of locating a place exactly. A person may know perfectly well that there are excellent ski slopes in Colorado, warm winters in Florida, splendid national parks and open mountain scenery in Wyoming, and that Boston is in Massachusetts. And such images may well determine a person's attraction to these areas. Many people drive to places just using signposts or national route numbers, or fly to them with complete faith in the airlines, without being able to locate them on a map. Many airlines provide booklets of maps for their passengers, but if you fly, take a casual glance down the aisles and see who is looking at them. The only people who seem to find them interesting are children – either because they are bored, or because they have yet to lose their natural curiosity about where they are.

Many adults apparently could not care less! Not long ago, a British journalist recorded this conversation between two young women in London:

'Where did you go for a holiday this year?'
'Oh, we went to Majorca.'

'Was it nice?'
'Absolutely smashing!'
'Where is Majorca?'
'I don't know exactly, I flew.'

Nevertheless, this young woman obviously had a glowing image
f Majorca, and presumably held one before taking off in a machine
1at transported her through a tube of ignorance between two points
n the Earth's surface. Furthermore, such attitudes are becoming
1ore common as more people travel by air and as more large auto-
vays are completed. Repugnant though the thought may be to a
artographer, some people may feel they know a great deal about
laces such as Bermuda or Alaska without having much idea about
1eir size or location – just as a person may admire a painting without
eing able to locate it in a particular time period or remember who
ainted it.

HE IGNORANCE SURFACES OF NORTH DAKOTA, LLINOIS AND PENNSYLVANIA

'or our first, and to geographers, discouraging, plunge into spatial
gnorance, we shall examine the situation in North Dakota (Fig. 4.1).
Jniversity students were asked to record the names of states on an
utline map, and by recording the proportion of errors we can draw
ontour lines enclosing areas of equal misidentification. Most striking
•n the ignorance surface are the high peaks over the eastern seaboard
vhere the states were misidentified, or not identified at all, over 60 per
ent of the time. In contrast, everyone knew they were in North
Dakota (geographers become increasingly grateful for such small
nercies these days), and with quite splendid perspicacity they all
guessed that the state just below North Dakota on the map was their
ister state of *South* Dakota. But from there on things became increas-
ngly grim. The ignorance surface rises away from the Dakotas, with
he marked exceptions of peripheral states with distinctive shapes, such
.s California, Texas and Florida. There is also a clear bias towards the
Vest, for the ignorance surface is generally low for all the northwestern and Pacific states. We will see why in a minute.

In northern Illinois Burris asked residents of Evanston, a town on the
1orthern outskirts of Chicago, to take part in a similar identification
•xperiment (Fig. 4.2). Unfortunately they did not fare much better
han the students in North Dakota. Illinois, with the peripheral states
•f California, Texas and Florida, is at the zero ignorance level; but
•eaks appear on the surface again in the East and in New England,
vhile Wyoming and the Great Plains states are only slightly lower hills

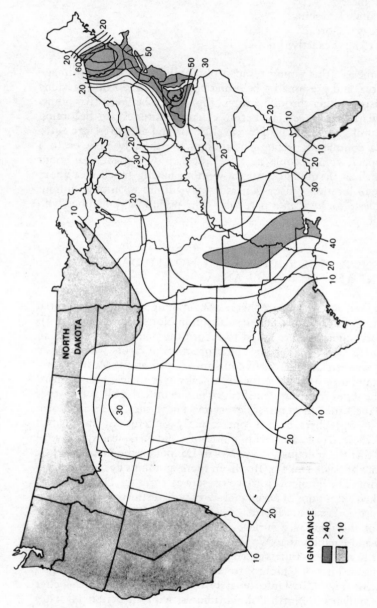

Figure 4.1   The ignorance surface from North Dakota.

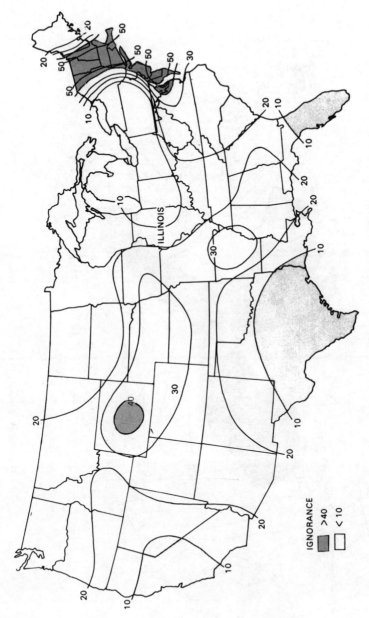

Figure 4.2   The ignorance surface from Illinois.

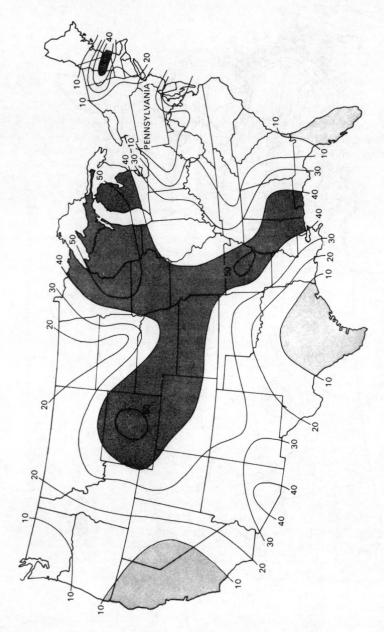

Figure 4.3 The ignorance surface from Pennsylvania.

of misidentification. The view from Pennsylvania is almost the same (Fig. 4.3), except that Pennsylvania is now at zero level, and large areas of the Upper Midwest, the South and the Mountain States are above the 40 iso-ignorans. Generally there seems to be a tendency for ignorance to rise with distance, unless a state has a very distinctive shape on the periphery of the national space.

This idea is clarified if we subtract the Illinois ignorance surface (Fig. 4.2) from that of Pennsylvania (Fig. 4.3), and map the differences to indicate where an Illinoian's knowledge is superior to a Pennsylvanian's, and vice versa (Fig. 4.4). Close to Illinois the students are much less ignorant than their peers in Pennsylvania, and all over the Far West they tend to be slightly superior, with the few exceptions of Washington, Oregon and Nevada. In California they are equally knowledgeable, while in South Dakota both groups are equally ignorant. However, in the East the Pennsylvanians are much superior, although their ability to identify states also declines with distance from their home area. The zero line, where the two fields of ignorance are exactly equal, lies, just as we would expect, about halfway between them.

Accounting for these surfaces of ignorance is not an easy task, for many things seem to be involved in varying degrees. A *distance* effect seems to be present, for in each case there is a tendency for ignorance to rise with distance away from the home state. But there are also a number of fairly consistent exceptions to this rule. Certainly *peripheral location* and *distinctive shape* help people to identify Florida, Texas, California and Maine, and only a small percentage misidentify a big state like New York.

THE CONFUSION MATRIX

People also tend to confuse neighbouring states with rather similar shapes, and we can record their mistakes in a 'confusion matrix', in which each row records the number of times a state was mixed up with another one recorded in a column. Occasionally we come across a person who thinks that Georgia, in the South-East, is really Washington, 3500 km to the North-West, but fortunately such spatial aberrations are rare. For example, when we map the confusion of Pennsylvanians (Fig. 4.5), we see that many of the mistakes are with neighbouring states. Vermont and New Hampshire are frequently mixed up, as are Arizona and New Mexico, Mississippi and Alabama, Illinois and Indiana, and the triumvirate of the upper Midwest – Minnesota, Wisconsin and Michigan. In all charity the original ignorance surface does seem to overstate the case: people confuse adjacent states sometimes, but they often know roughly where they are.

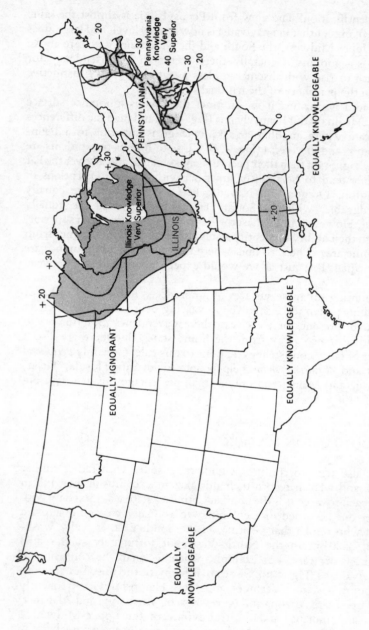

Figure 4.4 The differences between the ignorance surfaces of Illinois and Pennsylvania.

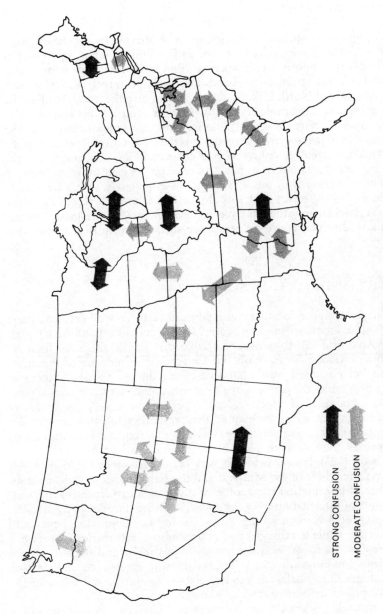

STRONG CONFUSION

MODERATE CONFUSION

Figure 4.5   The confusion map of Pennsylvanians.

THE TRAVEL FIELD

An ignorance surface is also going to be affected by the information that people have: indeed, in one rather limited sense, ignorance is simply the opposite of information. Presumably levels of information will rise, and ignorance will decrease, as people travel. We can see this in the case of North Dakota, where we can map the 'travel field' of the students in the sample (Fig. 4.6). As you compare this map to their ignorance surface (Fig. 4.1), notice how the ignorance rises as the travel field falls – particularly towards the East. Only 10 per cent of the students have ever visited the east coast and New England where the ignorance surface reaches peaks of over 60 per cent misidentification. The exception is in the West, but this is precisely where the steady decline of the travel field with distance is distorted. According to the travel field many students have visited Montana, Washington, Oregon and California, and their ignorance surface is distorted in the same way.

THE INFORMATION SURFACE

Shape, peripheralness, travel fields and distance – all seem to affect levels of ignorance in some degree. But what of information itself, that elusive, difficult-to-measure variable that might also shed light on these surfaces? We have obtained a very crude measure of the information possessed by the students from Illinois by asking them to record, under the pressure of a time constraint, all the towns and cities they can recall. It does, indeed appear to be a very naïve sort of measure, but when we map all their responses together as an information surface (Fig. 4.7), a very clear and interpretable configuration appears. A very high peak of information is located directly over Illinois (623), but the field falls very rapidly to somewhat lower levels over the whole of the Midwest and the East. Notice that the contour lines of information rise geometrically in order to represent the wide range of information over the country. Other, but distinctly smaller, peaks of information appear in California, Colorado, Texas and Florida, while the troughs and valleys of low information lie between these nodes in the South, through Appalachia, and over the Great Plains and mountain states. The relationship between the information and ignorance surface is moderately strong (0.54), and many of the exceptions are obvious and easy to explain.

But what is even more pertinent and intriguing is that the information surface itself can be predicted from two very simple variables that we considered before, when we were calculating population potential as an overall measure of accessibility. You will recall that the

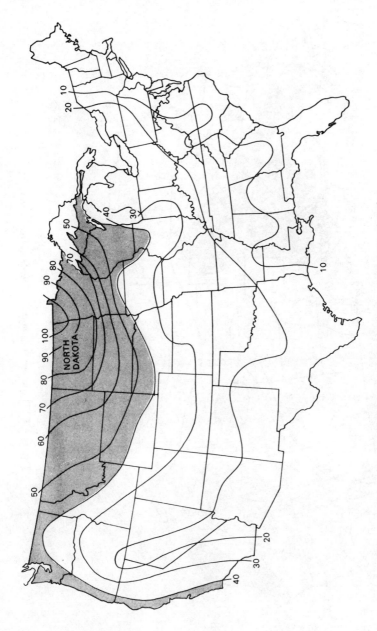

Figure 4.6 The travel field of students from North Dakota.

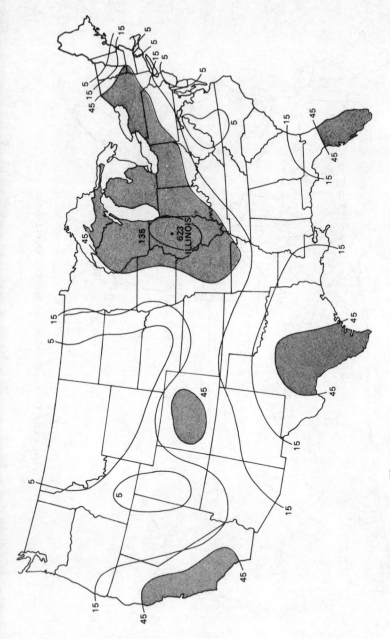

Figure 4.7 The information surface of students from Illinois.

population potential of a point involved both the distribution of the population in the area *and* the distance from the point whose access was being determined. If we now relate the information that people in Illinois have about other states, to the populations of the states, and the distances they are from Illinois, we can write:

information = f (population, distance)

Our guess is that as the population of a state increases we would expect the information it generates to go up, while the further away it is, the less information the people in Illinois will have about it. You can think of each state as a transmitter of information, and the people in Illinois as information receivers. After all, people generate information, and we might expect states with very large populations to generate very strong information signals that are gradually attenuated with distance away from the transmitting source. And this is precisely the case. After we have taken the logarithms of information, population and distance from Illinois, we can write:

log information = $-1.38 + 0.87$ log population -0.40 log distance

Remembering that adding logs is the same as multiplying the original numbers, subtracting them is the same as dividing, and multiplying logs by a constant is the same as raising the original number to that power, we have:

$$I = 0.24 \, \frac{P^{0.87}}{D^{0.40}}$$

There is a very strong and significant relationship of information to both these predictive variables. The equation fits our data quite closely (0.76 on a scale from 0 to 1), so that our ability to predict the information people have is surprisingly good once we know the size of the state and how far it lies away from Illinois.

Intuitive support for this numerical interpretation lies in a consideration of the alternatives open to a person faced with the task of writing down spatial information (place names) in a limited time. She has two basic strategies: to write down either the names of the largest and best-known places (Boston, San Francisco, etc.) or the names of the nearest places (Fig. 4.8). In actual practice most people use both strategies to produce an aggregate information surface that reflects both size and distance, and their relative location in information space.

From this very simple analysis, using our crude surrogate measures of information, we can now postulate an intriguing idea: the average person's information about geographical space is virtually determined

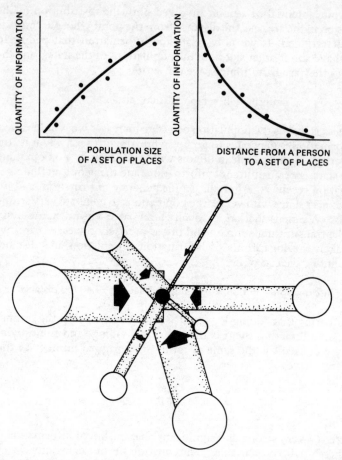

Figure 4.8 The quantity of information generated by places varying with the size of the places, and the distance they are from the perception point.

by his location within an invisible, but very real *information environment*. Given the knowledge of Mr or Ms Average Perceiver's location, and the distribution of people, we should be able to predict quite accurately the information they have in their heads about the space around them. This is an idea worth following up in another setting.

INFORMATION IN ONTARIO AND QUEBEC, CANADA

Our idea that information may be highly predictable from simple variables such as population and distance gains considerable support

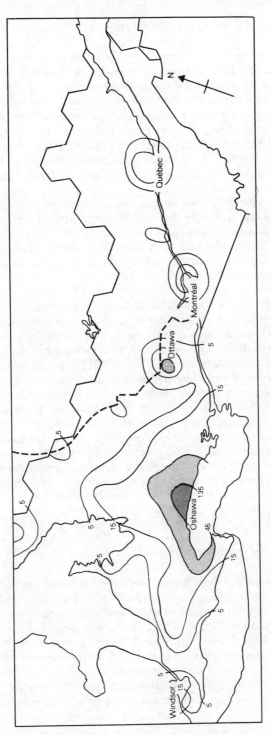

Figure 4.9 The information surface of 17-year-olds at Oshawa.

when we turn again to the French- and English-speaking teenagers in Canada (Ch. 3). For example, 17-year-olds in Oshawa generate an extremely sharp peak of information immediately over their home town which falls off quite smoothly with distance, except for small hills over major population centres such as Windsor, Ottawa, Montreal and Quebec (Fig. 4.9). Once again, there is a strong, significant relationship between information, and population and distance (0.77 on our 0 to 1 scale). This means we can predict with considerable accuracy the information these young people will have, simply from their locations.

The idea also helps us to understand some of the biases that French-speaking teenagers had in Montreal (Fig. 4.10, and recall Fig. 3.17). Another sharp peak of information appears, this time centred on Montreal itself, with smooth gradients in all directions away from the city – with the exception, once again, of such large population centres as Quebec City, Ottawa and Toronto. As usual, we can predict the information surface quite well (0.74), and it would be very interesting if we could examine the same information surface for the English-speaking 15-year-olds at nearby Pointe-Claire (Fig. 3.18). Unfortunately we do not have this, but such a comparison would raise the question of the way linguistic and other cultural effects shape these collective images of information. This would be an extremely difficult question to answer for technical reasons which we cannot explore here, but our conjecture is that these French- and English-speaking teenagers are exposed to very different streams of information. There are French and English language newspapers in Montreal, and people can tune in to either English or French television programmes. Moreover, vacations and visits to relatives may vary considerably between the two groups, reinforcing the directional and cultural bias we have seen. We shall explore these relationships between patterns of travel and information as we shift our attention away from North America and back to Europe.

INFORMATION IN SWEDEN

For our third look at these highly predictable information surfaces we will turn to Jönköping, a town in south-central Sweden, and look at the way in which the children there gradually acquire information about their country (Gould 1972a). Without implying any disrespect, we shall think of them as little 'black boxes' sitting in the middle of Sweden receiving information signals and storing some of them away for future recall. To get at this geographical information, the children were asked to 'spill' their information onto paper by writing down all the places they could think of under a time pressure – like the university

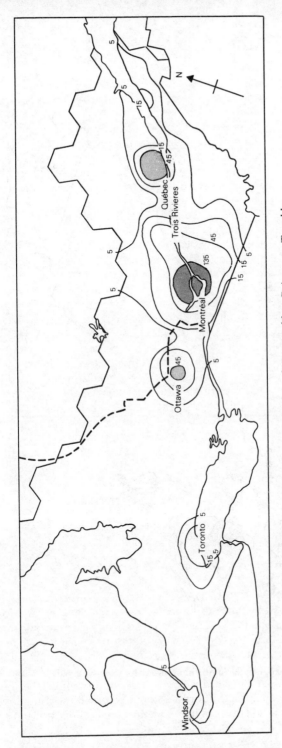

Figure 4.10 The information surface of 17-year-olds at Pointe-aux-Trembles.

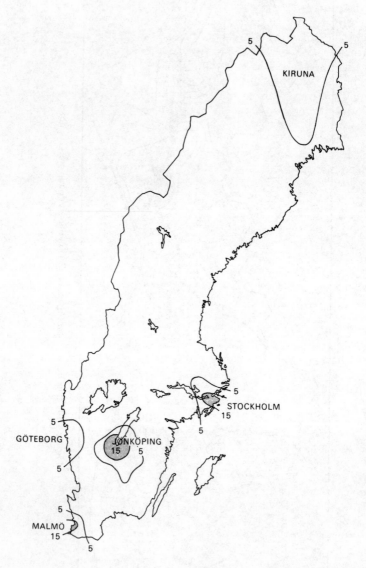

Figure 4.11 Jönköping: the information surface for seven-year-olds.

INFORMATION IN SWEDEN

students from Illinois. When the children are only seven years old (Fig. 4.11), the information surface is at a very low level. Again, notice that the contour lines are at the same geometric intervals to cover the range of variation, and that the surface is quite peaked. Immediately around the town of Jönköping there is a strong dome of local information, but elsewhere the surface is at zero level, or only just above it. The exceptions are three subsidiary peaks of information at Stockholm, Göteborg and Malmö, the three largest cities of Sweden forming a basic urban triangle in the South. There is also a small mound in the North at Kiruna, the big iron-ore mining centre, but this anomaly may be due to a strike in the mines at the time the sample was taken. Much to everyone's embarrassment, the strike was the first major labour dispute in Sweden for over a quarter of a century, and it received great publicity on all the television programmes.

By the time the children are nine years old (Fig. 4.12), the information surface has risen considerably. Again, the local peak is by far the most prominent, but the three major urban areas are still quite distinctive. Between them and Jönköping a filling-in process seems to be occurring, and a second hump of information appears to the north-east around large towns such as Örebro (Ö) and Linköping (L). Children also have a fair amount of information about the long island of Öland, one of the major holiday places for families in southern Sweden. The North is still a desert of information, with only a small blister appearing at Umeå, the capital of the North.

The information surface is strongly uplifted between the ages of 9 and 11 (Fig. 4.13), and the filling-in process proceeds apace between the high local peak and the major centres. The children are even acquiring some knowledge about the North, and what was previously a small blister around Umeå is now quite a respectable hill of information. Finally, at the age of 13 (Fig. 4.14), the surface has arisen and filled in again, but it looks as though some sort of saturation level is being reached. Nevertheless, the map appears to be a quite logical development from the three earlier ages. It is as though we are watching a photographic plate developing gradually over time. The image of the 13-year-olds, with good resolution, can be traced back as a complex development of the much simpler and more primitive surfaces of the seven- and nine-year-olds. First, small peaks of information push up around the local area and the large transmitting centres of the major cities, and then, gradually, corridors of information seep out from these nodes and link up between them. With this basic structure of nodes and links established, the surface continues to rise quite rapidly as the filling-in process proceeds.

Such a sequence of information acquisition is supported by the work of the geographer Douglass (1969) who worked with children in Illinois. She noticed that young children were able to specify a

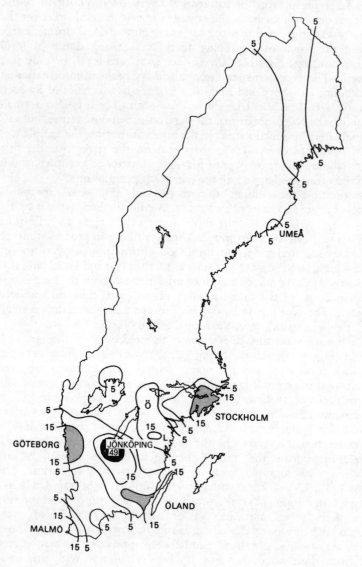

Figure 4.12 Jönköping: the information surface for nine-year-olds.

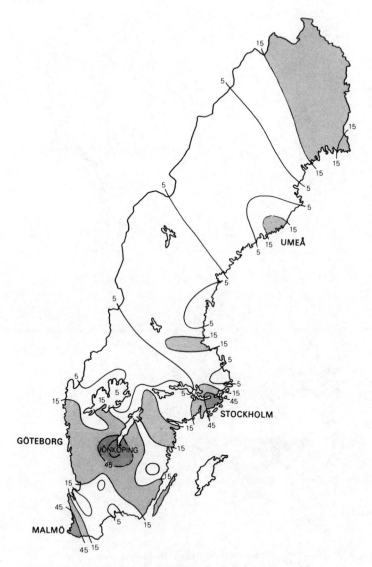

Figure 4.13 Jönköping: the information surface for 11-year-olds.

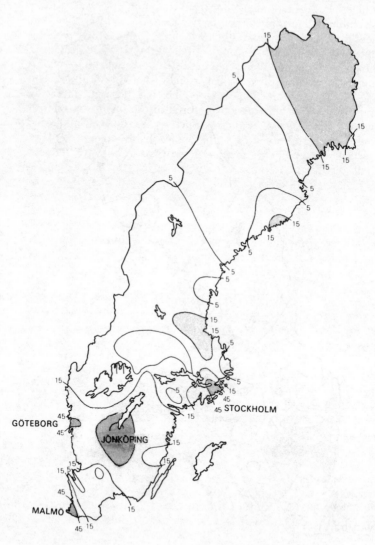

Figure 4.14 Jönköping: the information surface for 13-year-olds.

preference for certain large states, and there appeared to be a corridor of information to Florida, the premier vacation state in the East. This is exactly like the information corridor from Jönköping to Öland for the Swedish children. The information map sequence also closely fits some extremely imaginative and thought-provoking work of the psychologist Kaplan (1972), who works on the brain at the physiological level, and who has investigated those cognitive structures that allow people to acquire, store and process information so that they can adapt to changing environmental circumstances. He has postulated that an enormous amount of information is stored in the form of cognitive maps with distinct spatial properties such as location, distance and direction. These cognitive representations probably consist of major nodes, or reference points, joined by networks and linkages between them. His discussions remind one vividly of the developing information surfaces of Swedish children we have just examined, and appear closely aligned to ideas postulated by Lynch (1960) in his studies of human perception in American cities.

We can really move in two directions from the 13-year-old information surface at Jönköping: either forwards to the spatial preferences, or mental map of the children; or back to the things that might lie behind the information surface itself. There is no question that the information surface is closely related to their mental map (Fig. 4.15), but the measure of relationship (0.54), is lowered by several anomalies that point up the crudity of our information measure, and raise the very difficult question of trying to distinguish between positive and negative information, i.e. of putting a plus or minus sign on it. For example, the children have a fair amount of information about northern Sweden by the time they are 13, but the information obviously is not very bright because their preferences for the North are so low. Conversely, there are some areas in central Sweden where the children do not have much information, but the areas seem to acquire an aura of desirability from surrounding regions, and so have quite respectable scores on the mental map. When 10 (out of 70) such areas are removed from our analysis, the overall relationship between information and preference rises considerably (0.75).

We can also see how information surfaces are shaped not just by reading newspapers, journals and books, or by watching television, but by actual experiences of travel that expose people directly to information about other places. For example, adults at Jönköping (Fig. 4.16) have a very distinct 'travel field' when we plot the visits that a sample of them made for periods of one week or more. Once again, the visits seem to vary considerably, with many going to the larger cities and the pleasant vacation spots along the south coast (Gould 1975). Notice that only 9 per cent were made to the North of Sweden, reinforcing the marked lack of information that their children had of

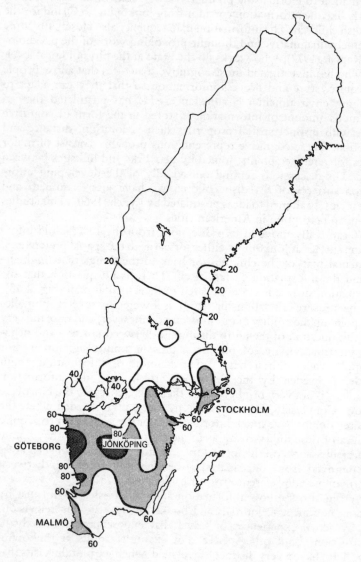

Figure 4.15 The mental map of 13-year-olds at Jönköping.

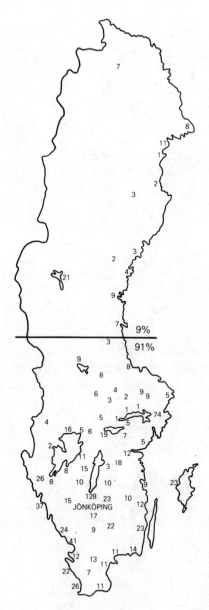

Figure 4.16 Visits of one week or more made by 128 adults at Jönköping.

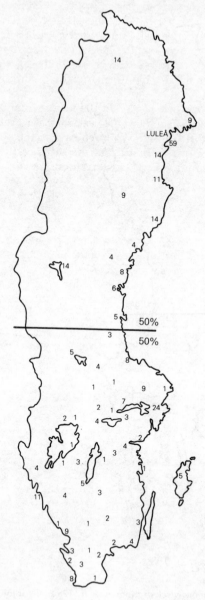

Figure 4.17 Visits of one week or more made by 59 adults at Luleå.

that part of the country (recall Figs 4.9–12). The travel patterns of people in southern Sweden are in marked contrast to the ones of those who live in the North. For example, the visiting pattern of people living in Luleå (Fig. 4.17) is split equally between the northern and southern parts of the country, which means that we might expect to find distinct directional biases in the travel behaviour of a country, with corresponding biases in the sort of information that is directly acquired by people.

Moving the other way, we can see if the information surface is as predictable from population and distance variables as it was in Illinois. Again, relating the information the children have to the population of the information generating regions, and the distance they lie away from Jönköping, we find a very close relationship (0.74), and our equation is now:

$$I = 3.98 \frac{P^{0.84}}{D^{0.41}}$$

Notice that the values measuring the effects of population and distance are almost the same as those we had for Illinois at a much larger geographical scale. There seems to be something extremely consistent about the information people have and their location in information space. At one time, our guess was that we might in fact have an expression of the form:

$$I = K \frac{P^1}{\sqrt{D}}$$

In other words, the information that people have about a region of their country is determined to a large extent by their location within that country; it is directly proportional to the size of the region's population, and inversely proportional to the square root of the distance away from their location. Unfortunately, we know now that these relationships are much more complex than this (Curry 1970). Indeed, they raise formidable theoretical and technical questions still being explored in advanced research (Fotheringham 1981). Most information surfaces *are* highly predictable, but there is no way yet of clearly sorting out the many different effects.

PEOPLE IN INFORMATION SPACE

Nevertheless, the fact that information is highly predictable from a person's location indicates that geographers have made an extremely important discovery here. These results indicate that people exist in what the psychologist David Stea (1967) has called 'invisible land-scapes', which shape quite strongly the mental images and the behaviour. We are reminded, once again (Chapter 1), of Herbert Simon's ant crossing a beach: her task is quite simple (to minimise her path to her nest), but her observed behaviour seems very complex because of the complex ripples and waves in her environment. To what extent does human behaviour seem complex because of the complexity of the information environment in which men and women are embedded? Certainly, in cities of the United States people's spatial behaviour is shaped by the hills and valleys of the invisible information and environmental stress surfaces over them. One French newspaper, *L'Aurore*, even published a map of Manhattan in New York City, indicating what areas were safe to walk in during the day and night. The *New York Times* (1971), immediately responded with a similar cartographic guide to Paris, but both were making the same point: invisible stress surfaces lie over both cities, and influence people's paths and movements to a considerable extent. In Sweden a study in educational psychology has shown that scores on a standard battery of intelligence tests are significantly related to simple measures of access, based, in turn, on population potential notions. Children in areas of high accessibility, literally in the swing of things, tend to have a small but significant advantage on certain tests, owing to their locations in these intense nodes of human activity and communication. Interest-ingly, the urban–rural gap seems to be shrinking, as information flows are quickening and gradually wiping out former and well established disparities. Even so, location in information space appears to have some measurable effect upon human abilities.

People's preferences for places seem to depend upon their location: to what extent does this idea carry over to other areas of human preference, and the way such likes and dislikes are expressed by actual behaviour – towards other people, to other areas, countries, cultures, religions and so on? An older geography concentrated upon an environmental determinism in the physical realm, but became so platitudinous that it was rejected by a later generation. Perhaps geographers should focus instead upon the invisible human landscapes which seem to be far more relevant, both for understanding human spatial behaviour, and planning a better spatial environment for people in the future.

BARRIERS TO INFORMATION FLOWS

In the examples from Illinois and Jönköping, we have looked at situations where there are very free flows of information over geo-graphical space. Both groups of students were asked about their own

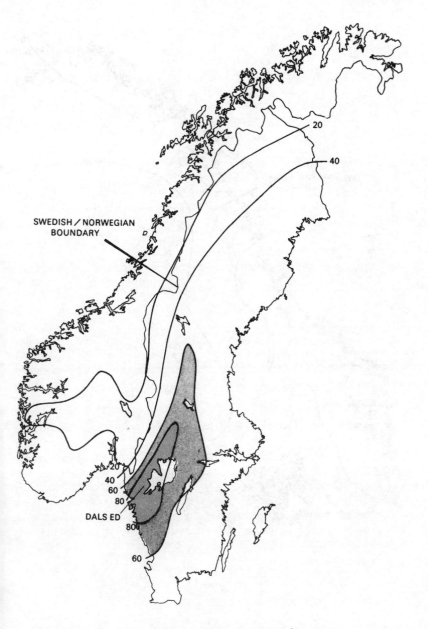

Figure 4.18 The mental map of schoolchildren at Dals Ed, Sweden.

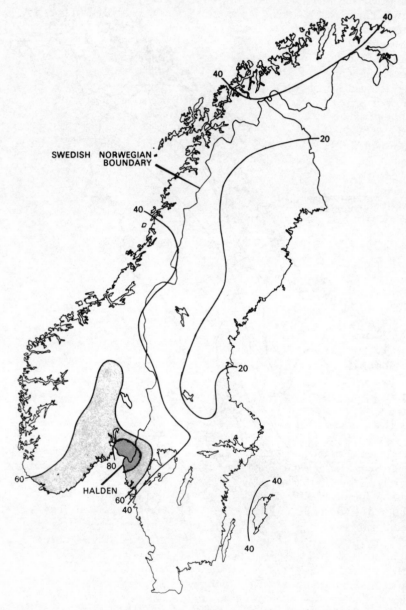

SWEDISH NORWEGIAN BOUNDARY

40

40

20

40

20

60

80

HALDEN

60

40

40

40

40

Figure 4.19 The mental map of schoolchildren at Halden, Norway.

ountries, and there were no discernible barriers to the information ows transmitted to them from the various regions – except the ttenuating barrier effects of distance itself. But obviously other arriers may exist, and they could take on a variety of forms. Some are ormed by natural features, such as mountains, long lakes and so on. These might have the effect of cutting down information flows, just as hey frequently make other forms of human interaction more difficult. Most of us are also well aware of linguistic barriers: our information in foreign country, whose language we do not speak well, is severely educed by the linguistic filter. Political boundaries may also affect the nformation people have, as we shall see from this comparison of the reference surfaces and information fields of children in Norway and Sweden in schools only a few kilometres apart, but on either side of the oundary between the two countries.

On the Swedish side, the pupils at Dals Ed display a marked referential bias toward locations in Sweden (Fig. 4.18). There is the usual strong dome of desirability centred over the perception point, nd the surface drops quite smoothly away with distance except for a maller peak at the island of Visby – a very attractive and deservedly amous holiday area. But notice how asymmetrical the surface is: the order with Norway is analogous to a geological fault line, for on the Norwegian side the surface has slipped down to much lower levels.

When the perception point moves just a few kilometres over the order, to the town of Halden in Norway, a very different perception surface characterises the preferences of the pupils (Fig. 4.19). Although the local dome is somewhat lower, and spreads slightly into Sweden, there is still a distinct bias towards Norwegian locations. Again, the immediate area is highly valued, together with the beautiful country across the Oslo Fjord. There also seems to be a distinct aversion towards cities, for not only are the cities in Sweden given low perception scores, but nearby Oslo is not regarded very highly. We are reminded of the attitudes of many pupils in southern England towards their own Metropolitan Sinkhole, although in all justice Oslo could hardly be described as such.

The border also strongly warps the information fields of the children. At Dals Ed in Sweden (Fig. 4.20), the children can readily recall names of places in their own country, but very few in Norway. The information field of the Norwegian children at Halden is just the opposite (Fig. 4.21). Clearly the boundary has a very severe barrier effect upon the flows of information to children only a few kilometres apart in geographical space. Although the language differences are slight, the radio and television programmes expose them to different sources of information. The Swedish geographer Lundén (1971) has also shown how the geographical content of standard school books varies markedly, always giving the children more information about

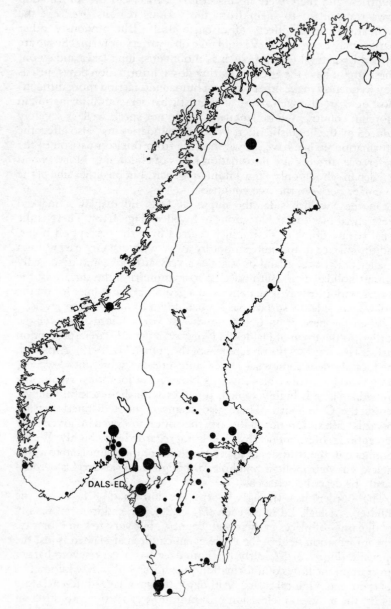

Figure 4.20 The information field for children at Dals Ed, Sweden.

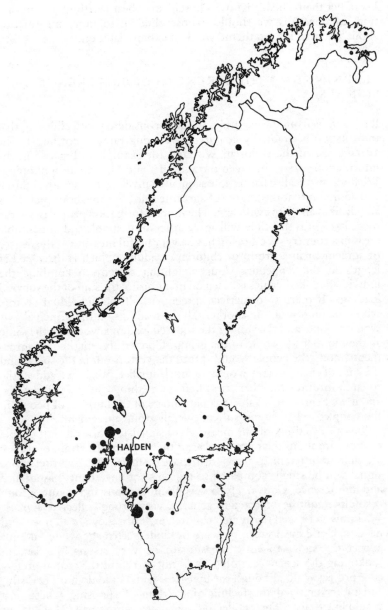

Figure 4.21 The information field for children at Halden, Norway.

Us rather than Them. 'Know Thyself' may be a worthwhile admonition, but perhaps we should expose children to more information about Them. This is a theme we shall explore later on.

LEARNING AND THE STRENGTH OF PREFERENCE SIGNALS

It is obvious from the examples we have considered with children, that information is gradually acquired at varying rates, according to the average age of the children. We saw, for example, a big rise in the information surface between the ages of 7 and 11 years in Jönköping. What we also realise from these sequential investigations with children and adults, is that the degree of agreement in a group about their likes and dislikes also rises with age. The more information people have, the more likely it is that they will agree upon the desirable and undesirable areas in a country. We already have a very useful measure of the degree of agreement in a group of children or adults, which is discussed at length in the Appendix. Our weighting system to combine the individual ranked values is based on maximising the sum of the squared loadings. If there were perfect agreement, all the individual vectors would lie on top of one another, all the loadings on the principal axis would be one, and the sum of the squared loadings would be the same as the number of people in the group. Conversely, marked disagreement among the people would spread the vectors out in the space, and the sum of their squares would be much smaller. When we add all the squared loadings together, we obtain something called an *eigenvalue*, and if we express this value as a percentage of the number of people in the sample we have a measure of overall group agreement.

We can also think of our measure as a signal-to-noise ratio. If people are not sure about their preferences, if they lack information, or if they are quite indifferent to many of the areas, their collective preference signal will be small, compared to the noise of uncertainty, ignorance and indifference. Conversely, a very homogeneous and mature group of adults might be expected to generate very strong collective signals. We know very well that when children are born they are all noise and no signal, and even when they are of kindergarten age we would not expect the signal-to-noise measure to be very strong. In fact, in Jönköping the seven-year-olds could not carry out the spatial preference project at all, although the nine-year-olds could do it quite easily, and rather enjoyed the feeling of freedom of choosing where *they* would like to live. Their collective signal was quite small (24.0%), but significantly different from one that might have been generated by just random preference rankings (7.1%).

As the children become older (Fig. 4.22) their preference signals rise

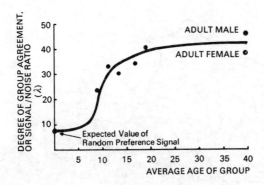

Figure 4.22 The relationship between group agreement (λ) and age in Jönköping, Sweden.

very quickly at first as they acquire and store away large amounts of information. We recall how rapidly the information surfaces rose between the ages of 7 and 11 (Figs 4.11–4.14). But at 13 the 'learning' curve begins to turn over, and finally levels out at 18 and adulthood. What we have seen emerge is an increasingly strong preference signal with age, described by a curve that is identical to many others that record learning situations. Until children are about seven or eight their preference signals are so small that we cannot really detect them against the background noise; but then a sudden flood of information impinges upon them, and the preference surfaces begin to crystallise out of the mental flux. By adulthood the surfaces have set pretty firmly with a high degree of agreement about them.

THE GROWTH OF SPATIAL PREFERENCES IN WESTERN NIGERIA

To see how preferences themselves change with age, we are going to turn to West Africa for our final example. The former Western Region of Nigeria is structured by a crescent-shaped core area (Fig. 4.23), containing nearly all the large towns. High road densities stretch from the federal capital at Lagos through the large cities of Ibadan and Ado-Ekiti, to Ikare in the east. We are going to look at the mental maps of children and young adults at Oyo, which lies just to the north of Ibadan at the edge of the core area (Gould & Ola 1970).

When the pupils are 13 years old (Fig. 4.24), there is a very strong local effect, with corridors of high preference reaching out to Shaki in the north-west and Ilesha in the east. The only other area to score above the 60 isoline (shaded) is Lagos to the south. However, two years later

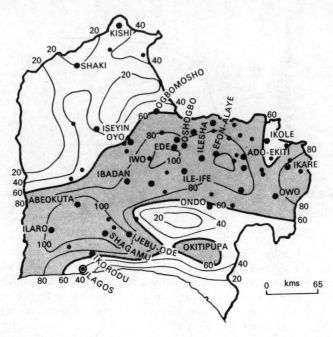

Figure 4.23 Western Region, Nigeria: the crescent-shaped core area of high population and road densities.

the very strong local effect is starting to dissolve (Fig. 4.25), as preferences are assigned to areas in the core region. Like an amoeba, the local effect has 'oozed' eastwards, breaking the previous long corridor of preference to the north-west. Now towns such as Abeokuta, Ijebu-Ode and Akure form smaller preference domes. By the time many of the pupils are about to leave school at 18 (Fig. 4.26), these local domes have themselves connected up, and five years later the preference surface of student teachers has consolidated into two major ridges and an outlier, all within the core area of the region (Fig. 4.27). If we subtract the 13-year-old surface from the 23-year-old (Fig. 4.28), we get a very clear expression of the changes in preference that have taken place. All the positive changes lie directly along the crescent-shaped magistral, which contains the greatest information-generating capacity with its many people travelling over a dense and relatively well connected transportation system. Conversely, the 'bush' areas to the north-west and south-east are all negative, implying that early preferences of the younger pupils have been removed and assigned to areas more desirable to adults.

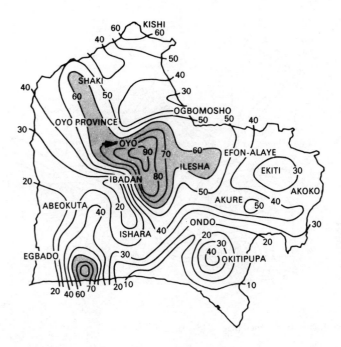

Figure 4.24 Oyo: mental map of 13-year-olds.

Clearly, as the pupils grow older, they acquire greater information about those parts of their region that generate large quantities of it. The information surface affects in turn their preferences, and gradually pulls the perception surface into conformity with it. The information also affects the degree of agreement in these groups (Fig. 4.29). While we do not have measures for the younger ages and must make some reasonable guesses, it is clear that we have an agreement function that closely resembles a learning curve again. The upper line for Oyo, the town lying on the edge of the core region and well connected to it by a federal trunk highway, rises very quickly between the ages of 8 and 13, and then slowly converges upon an upper level around 55. In marked contrast, the agreement curve for pupils of similar ages in Okitipupa, a much smaller town in the south-east, without any trunk highway connection, rises much more slowly, and seems to level off at a lower value of agreement. Once again, we have evidence that location in the information space has a distinct and measurable effect upon people's information, preferences and behaviour.

Thus, location in information space, a person's position in this one of many invisible landscapes, has emerged as a critical consideration if we

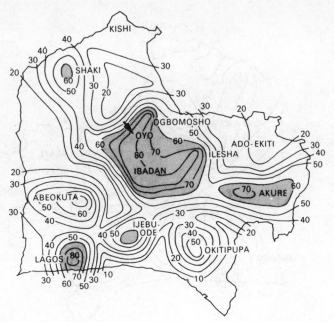

Figure 4.25 Oyo: mental map of 15-year-olds.

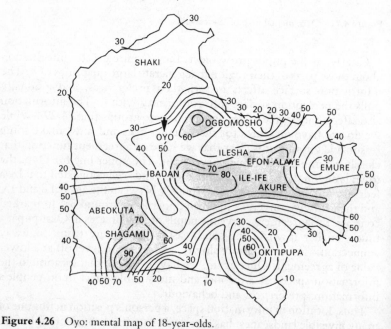

Figure 4.26 Oyo: mental map of 18-year-olds.

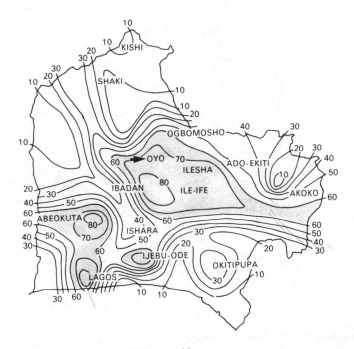

Figure 4.27   Oyo: mental map of 23-year-olds.

are to understand human behaviour in a spatial or geographical setting. Given the importance of relative location – that is, location relative to the information-transmitting sources in an area – we might expect a number of human decisions to be shaped by the varying intensities of information streams and the way they, in turn, shape people's perception of the geographical space around them.

In the next chapter, we have chosen just one area where a knowledge of these mental maps may be important. For this is the sort of knowledge that contributes not only to an understanding of others, but helps us to be more self-aware, as we realise that our mental images may be shaping our perception and evaluation of places, people and events. The area we have chosen is public administration, and the question that of assigning people to locations in geographical space in order to carry out some required tasks. To carry out a task well, people must feel happy and comfortable where they are. If they are not, perhaps some additional compensation may be appropriate. It is to these questions that we now turn.

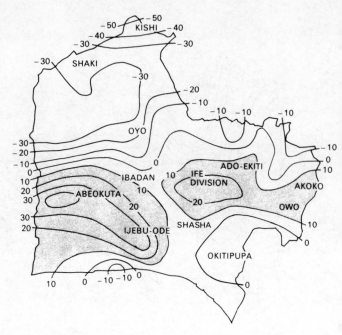

Figure 4.28 Oyo: changes in perception; 13- to 23-year-olds.

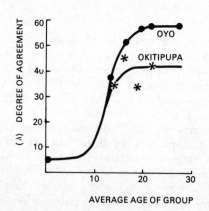

Figure 4.29 The relationship between group agreement (λ) and age in Oyo and Okitipupa, Western Region, Nigeria.

5 Mental maps and administration

'No duty the Executive had to perform was so trying
as to put the right man in the right Place.'

Thomas Jefferson in J. D. MacMaster,
History of the people of the United States

One of the most profound changes that has come about during the past 100 years is the incredible consolidation and centralisation of administrative control in many areas of public and private life. Whether we consider the ambassador removed many weeks, or even months, from his minister or king; the colonial officer far up-country from the governor; or the commercial representative negotiating on behalf of his firm at home – all had a freedom of initiative and areas of wide discretion usually denied their counterparts today. The difference, of course, is the explosive development of electronic communication and air transport. Too often today the negotiator or administrator is simply a lap dog at the end of a long electronic leash, and any attempt to deal with events 'not in the book' usually involves requests back to the seat of power and decision in the centralised headquarters of the nation, organisation or firm. It is little wonder that men and women of intelligence and initiative are increasingly frustrated by bureaucratic institutions that throttle them with cords of rapid communication.

ALLOCATING PEOPLE TO PLACES

Yet, despite our increasing ability to communicate and look over some poor devil's shoulder, there are many things which cannot be controlled effectively from the centre. An educational radio or television programme may be beamed over hundreds of thousands of square kilometres, but the child in the Australian bush still profits immensely from the face-to-face communication of a real teacher. Even with older people, the Open University in Britain has devoted an enormous amount of thought and effort to the task of maintaining direct contact between students and tutors. In any educational programme close communication between human beings, rather than with faceless bureaucracies, is essential. Illnesses may be diagnosed by computers and blood-fractionating machines at some centralised facility far from the patient in a small rural infirmary, but a trained doctor must be there to examine and to cure. Administrative directives setting general policies may gush from the offices in the capital, but if there are not men-on-the-spot, then the men-behind-the-desks might as well whistle in the wind. In a developing country, or in Iowa, agricultural policies require for their effective implementation the devoted, and too-seldom appreciated, work of the assistant agricultural officer and county agent far from the bright lights of the city. And despite videophone and teletype, large commercial contracts are often closed after intense face-to-face negotiations and bargaining in places remote from the metropolitan headquarters. Even sales are greatly affected by having highly trained personnel serving in out-of-the-way places – as Volkswagen so astutely realised in East Africa and other parts of the world.

All these cases raise the same fundamental question: despite our growing ability to communicate quickly and cheaply, we still need people on the spot. But how do we get men and women – teachers, doctors, administrators, engineers, dentists, technicians, salesmen, construction workers and so on – to serve in spots that may be perceived as undesirable? In brief, how do we allocate scarce personnel over geographical space, so that not just the hills of the perception surfaces are served, but the valleys as well? This is a very real problem facing both economically developed as well as developing countries. Allocative systems can, of course, be coercive, but they are hardly in favour. The disgruntled administrator longing for another area is unlikely to be effective; and the fed-up teacher probably does children more harm than good. A sense of duty may keep a person going at an undesirable place for a while, but underlying resentment may affect many areas of performance. The suppressed frustrations of the dutiful servant are not the strongest foundations upon which to build a true sense of service.

Every country, then, faces the problem of spatial allocation of scarce human resources, and every governing body wrestles with this difficult question. In the Soviet Union, very high incentive payments are made to skilled construction personnel working on large Siberian hydroelectric dams, in an attempt to draw men and women from the highly desired western areas of the country. Alaska pays teachers higher salaries, partly to overcome the high cost of living, and partly as an incentive payment to teach in winter environments that are not everybody's first choice. Australia pays a bounty to those willing to teach in the remote, hot outback stations; and huge incentive wages are paid by oil firms to the riggers working through the bitter Arctic nights of the Beaufort Sea.

But remoteness is partly a mental construct, and need not be defined simply in terms of the Arctic shore or the Australian bush. Many areas of the United States seem to be perceived as equally remote and undesirable. There is a chronic shortage of doctors, dentists and nurses throughout Appalachia, and we are seeing one pathetic case after another of a little community building a small and well equipped medical centre but being unable to attract a young doctor to it. In Sweden the same problem arises, and the remote and sparsely settled North exerts a continual demand upon the medical and dental professions. In Britain the geographer Davies (1966) has shown how the strong place preferences of new doctors simply reinforce the present pattern of surpluses and deficits. The strong desire to work in the South of England builds up a latent migration potential which may serve as an early-warning system of future movements. Nigeria faces exactly the same question, and the *Morning Post* of Apapa noted on 15 February 1972:

> Nigerian doctors will now be required to spend a year in the rural areas of the country before they could be placed fully on the register. This decision was taken by the Nigeria Medical Council at its last meeting. ... The registrar advised the various governments in the country to give financial and other incentives to doctors during their service in the rural areas.

It is hardly surprising that the American Peace Corps, the British Volunteer Service Overseas, and similar programmes of other countries, have enjoyed such marked success at the local people-to-people level – despite occasional contretemps on the political and diplomatic planes. The allocation of dedicated young men and women to all sorts of jobs in remote rural places has frequently had a catalytic effect on the people, particularly the children of the areas. Within Nigeria, a year's service with the national Youth Corps is compulsory for university graduates. They are sent outside their home state to serve in agriculture, education and health care. Some of the more remote states are now heavily dependent on the Corps to staff their vacant posts.

Unfortunately such dedication is not always displayed. That these spatial biases can have tragic consequences is illustrated by the situation in northern Chad from 1968 to 1970, as described by Jacques Derogy in *L'Express* of 24–30 August 1970:

> The 'administrators with glasses', who were promoted in 1960 in place of the traditional chiefs to turn this mixture of ethnic groups and old-fashioned methods into a modern, centralized republic, preferred the comfort of air-conditioned offices to life in the bush. They abandoned their posts, and left huge territories in a state of

anarchy. For the last year, no more than half the taxes have been paid. The cotton harvest has fallen by a third, and the export of livestock is disorganized. The discontent is spreading through the savanna of the south.

Ten years later a protracted civil war between the 'Forces of the North' and southern supporters of the central government led to the complete devastation of Chad.

GOVERNMENT SERVICE IN TANZANIA

The problem of getting well qualified people to serve in places that need them desperately, but are perceived as low valleys on a mental map, is illustrated by two examples from Africa – Tanzania and Ghana. The problem in Tanzania is especially critical, for the country is large and very unevenly populated, and there are very few university graduates available at the moment. Unfortunately, some of these appear to hold very superior views of themselves, and on occasion President Nyerere has had to deal firmly with student unrest whose underlying cause was an extreme élitist viewpoint. Such a view is all too common in many countries where the universities were originally set up by former colonial powers. Many quite unsuitable outer forms were adopted, such as gowns (in 35°C heat!), common rooms with port and madeira, and academic subjects quite useless to a developing country with its own rich cultural traditions. In Ghana, for example, classics were represented by Greek and Latin, but no Arabic studies were available. The practical land grant university tradition of the USA would have been much more relevant. Given the stupid attempts to transfer the medieval traditions of European institutions to totally inappropriate settings, it is hardly surprising that many university students in Africa acquire such attitudes.

The mental maps of the Tanzanian and Ghanaian university students were measured in two ways (Gould 1969, 1972b). First, they ordered their preferences in the now familiar way on a map, responding to the situation as though they were entering government service upon graduation, and had a choice of posting to a particular district in their country. Secondly, they evaluated each district on the following five scales measuring their perception of the area's cost of living, accessibility, environment, facilities, and the local people:

Salary

Expensive place to live in, saving from salary hard.	+3 +2 +1 0 −1 −2 −3	A very cheap place to live in, I can easily save from salary.

Travel
Very easy to get there, movement in or out no problem.

+3 +2 +1   0 −1 −2 −3

Very difficult to get to, and very hard to get in or out.

Surroundings
Very pleasant, an extremely nice place to live.

+3 +2 +1   0 −1 −2 −3

Very unpleasant, not a nice place to live in.

Facilities
Many things to do, spare time is no problem.

+3 +2 +1   0 −1 −2 −3

Not much to do in spare time, far from centre of things.

Local people
Friendly and very easy to get on with.

+3 +2 +1   0 −1 −2 −3

Very difficult for me to fit in and get on with people there.

The mental map of Tanzanian students (Fig. 5.1), virtually picked out all the major towns in the country as high peaks of preference, while the rural areas were held in very low esteem. The perception surface matches almost perfectly a map showing the degree of modernisation (Fig. 5.2), and it is clear that their desire to serve in areas with few modern facilities is extremely weak. While Dar es Salaam, the capital, is only moderately high, the bright lights of the cities generally seem to shine with great intensity in the minds of these students, drawing them towards the major nodes of administrative life. Arusha and Moshi, in the shadow of Mt Kilimanjaro, are a major peak, but other highland centres such as Morogoro, Iringa and Mbeya are also greatly desired. Smaller peaks appear at major administrative centres such as Tabora and Mwanza, and the lowest point on the entire surface is at Masasi in the far south, an area virtually disconnected from the rest of the country.

The responses of the students to each district on the different scales also confirmed these basic geographical preferences. The five scales are highly interrelated, indicating that if a district is seen as desirable or undesirable on one ruler it is very likely to be seen in the same way on the others. Such a tendency to collapse all the scales into one reminds us strongly of the students in Illinois trying to cope with their cognitive dissonance by combining all their scales to avoid perceptual conflict.

However, there are also some important secondary effects here. If we add all the students' responses together, we can obtain a score of each district on each of the five scales. The *people* scale is much less strongly related to the other four, and the students respond to this question in interesting ways. On this important ruler, measuring the degree to which university students feel they can fit in with their own

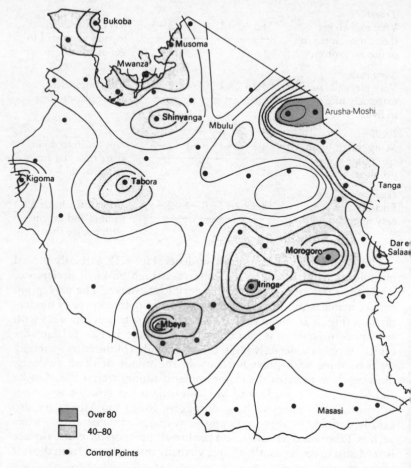

Over 80

40–80

• Control Points

Figure 5.1 The mental map of Tanzanian university students.

people when they are on government service, the average score of all the districts in Tanzania is only 3.6, and 37 per cent of the districts receive average scores that are negative, implying outright dislike for the people there. Many of these are along the coast where the Arab influence is strong, in areas where the Masai are dominant, and in the far western portion of the country.

GOVERNMENT SERVICE IN GHANA

In Ghana the responses of a group of students at Accra also pull out the major urban areas as high peaks on the mental map (Fig. 5.3), and there

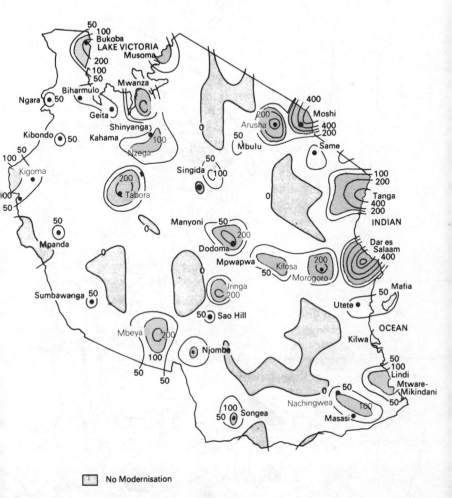

Figure 5.2 The Tanzanian 'modernisation surface'.

is a strong preference for the southern portion of the country. The capital Accra, the large port city of Takoradi and the major centre of Kumasi in Ashanti form the basic triangle of southern desirability with secondary peaks at Koforidua, Oda, Obuasi and Sunyani. The major exception is a high, outlying peak at Tamale, the capital town of the North, but perceptual gradients are very steep as one moves away from this isolated node. Steep gradients characterise the surface in general, for even in the South deep pockets of undesirable areas appear only short distances from major peaks. The shore of Lake Volta is also

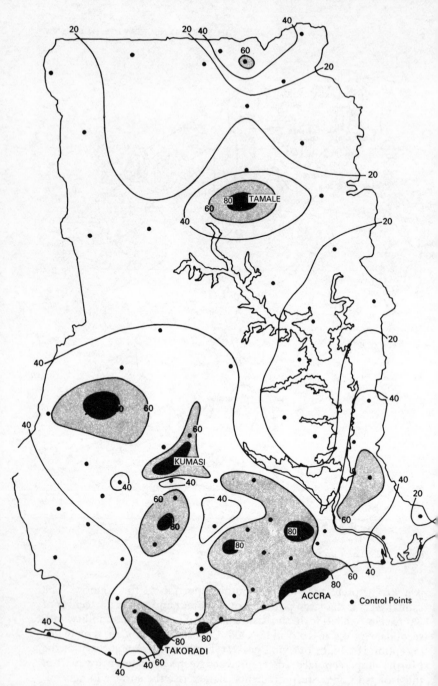

Figure 5.3 The mental map of Ghanaian university students.

shunned, for this massive man-made water body has severed many connections, and the area appears remote, inaccessible, and hardly a place to be posted on government service. At this general level, there seem to be few differences between the Ghanaian and Tanzanian students. Both groups long for the bright lights and bustle of the towns, and it is hardly surprising, if these élite groups like urban life, that many of their countrymen share their feelings. A strong tide of rural-to-urban migration characterises many African countries today, as people flock to these intense nodes of human life and communication where the important things seem to be happening. Unfortunately, many of the really crucial aspects of development should be taking place in the rural areas, for there must be considerable reliance upon agriculture, both for feeding the people, as well as providing the source of materials for an embryonic manufacturing sector of the economy.

It is when we examine the Ghanaian students' responses to the five scales that significant contrasts with their Tanzanian peers are highlighted. While the Ghanaians also collapse the scales to avoid cognitive dissonance, they again regard the *people* scale as rather different from the others. However, the average score of a district on the *people* scale is 19.6 (in contrast to the Tanzanian 3.6) and only 5 per cent of the districts receive quite small negative scores. We might feel, from this evidence, that compared to the Tanzanians the Ghanaian university students feel that they could fit in much more easily, and work quite positively with by far the greater majority of their fellow countrymen.

CHINESE AND MALAYAN VIEWS OF MALAYA

For our third, and final, example we are going to shift to South-East Asia, and look at the mental maps of two rather distinct groups in the same region (Leinbach 1972). Students in Malaya (Peninsula Malaysia) may either be of Malay or Chinese background, and while the region as a whole has a small Malay majority, the two groups are distributed very differently. This means that in some areas, particularly the rural ones, Malays may far outnumber their countrymen of Chinese background; while in other areas, especially those that are urbanised, the Chinese population dominates.

Alor Star, in the northwestern corner of the region, is in the heart of the most significant rice-growing area of Malaya. Both Chinese and Malay 18-year-olds were asked about their geographical preferences, as if they had a completely free choice of serving in any district of Malaya and Singapore after graduation. The mental map of the Malayan students (Fig. 5.4) displays an extremely strong local effect centred over the town of Alor Star itself, but this dome of local desirability extends a considerable way into the surrounding rural areas. These are all dominated by Malay farmers growing rice. The only other high peak is the island of Penang, a beautiful holiday area with white, palm-lined beaches and a decidedly slow and leisurely pace of living. Small and quite moderate peaks appear at Kuala Lumpur, the capital of Malaysia, and at small towns on the east coast such as Kota Bahru and Kuantan, which serve mainly as agricultural service centres for a largely Malayan rural population. Towns are obviously not that attractive to these Malayan students, and the predominantly Chinese city of Singapore is the zero point on the entire perception surface.

However, through Chinese eyes Malaya is viewed somewhat differently (Fig. 5.5). The highest peak on the mental map is not at Alor Star itself, but over the island of Penang and the Chinese-dominated commercial mainland adjacent to it. The local dome is much more constricted than in the case of the Malayan students, and obviously if an area is only a short distance from Alor Star its residential desirability is greatly jeopardised in Chinese eyes. To the south the Kinta valley in the heart of the tin-mining area of Malaya is perceived as high, and another large but very local peak is centred over the capital at Kuala Lumpur. On the coast the towns of Port Dixon and Malacca are picked out as secondary peaks, while to the south the island and city of Singapore is the second highest node on the entire surface. This huge, sprawling Chinese-dominated city is such a magnet that even the adjacent mainland of Malaya and the town of Johore Bahru are swamped by high-perception scores. The entire east coast is well below the average, and it is obvious from talking to the Chinese students that any posting there would be regarded as a definite hardship. On both the Chinese and Malay surfaces, of course, the rugged and uninhabited jungle area of Kelantan forms a very low trough of desirability.

Clearly, from this example, the perception of residential desirability is partly affected by sets of cultural attitudes that may differ in marked ways between groups of people of the same age in the same location. The Malay students obviously feel more comfortable serving in areas where their own people predominate, and they have much less desire to move away from the local areas where they were born and raised. Once away from Alor Star and its rice-growing countryside, only a few moderate peaks push up through the generally low-lying perception surface. The Chinese students, in marked contrast, view the whole western area of tin mining and rubber as desirable, and on this large plateau of perception there are high urban peaks. The greatest contrast, of course, is Singapore itself: the second highest peak for the Chinese, but the nadir of desirability for the Malays. A more vivid effect of cultural attitudes upon perception and mental maps could hardly be found.

Figure 5.4 The mental map of Malayan students at Alor Star.

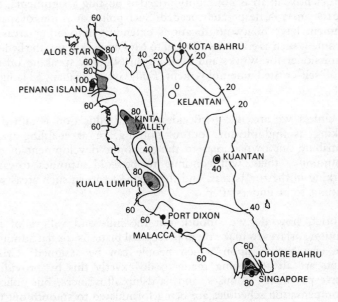

Figure 5.5 The mental map of Chinese students at Alor Star.

SMOOTHING MENTAL MAPS

In all these examples we have seen the way in which mental maps raise severe policy questions as these developing countries face the very difficult problem of assigning skilled men and women to serve in the valleys and troughs of the perception surfaces. What, for example, are the policy implications of the radically different Chinese and Malayan viewpoints? Does a government, desperately trying to create a unified nation out of such cultural diversity, assign Malays to Malays and Chinese to Chinese, to accentuate the differences? Does it deliberately assign administrators, civil servants, doctors and technical personnel to areas where the people have different cultural backgrounds, in an attempt to create a melting-pot? Or does it ignore the problem altogether? Such questions are extremely difficult to answer, yet they face most countries in one degree or another. In Tanzania, for example, the same question arises when a man or woman, say an assistant agricultural officer or a nurse, wants to be assigned to the home area, where the language and customs are comfortable and familiar. Are they more effective in the short run, working and serving in their own areas, or is the long-term planning horizon to a strong and united nation more important?

And apart from cultural differences within a nation, how does one formulate policies that will get people to serve, cheerfully, willingly and efficiently, in areas not highly prized as posting assignments, but where they may be desperately needed? Such policy questions of spatial assignment have many ramifications extending into other areas of government such as education. Listen to Mr Femi Okunnu, the Federal Commissioner for Works and Housing in Nigeria, speaking on the topic of 'self-created' unemployment (*Daily Sketch*, Ibadan, 28 January 1972):

> ... Unless we are able to decide whether education is aimed at making us find pleasure out of nothing, or at enabling us to contribute our own quotas to the economic development of our community, there will continue to be cold attitudes towards working in the rural areas, and the labor sectors in such areas will continue to be understaffed.

In brief, how do we smooth out the hills and valleys of the perception surfaces to make them perceptual plateaus, or flat administrative chessboards upon which people can be assigned? Crude attempts are already being made to do exactly this by providing incentive payments, bonuses, and hardship allowances, but policies and compensation schedules are being formulated to smooth out the peaks and pits on mental maps, without much knowledge of what is being smoothed. And at this policy level, individual differences, which we have largely ignored, come into play. Within a group of civil servants, for example, there may be some who greatly like certain areas shunned by the majority. For example, a personnel officer for the Nigerian Electrical Power Authority, who was looking for volunteers from the Benin office (in the south) to move to the Kano office (in the north), was pessimistic. However, a volunteer immediately came forward. The reason? His wife was convinced she would be able to do well in the fruit and vegetable market in Kano, 'importing' goods from the south to serve the southern community already established there.

Perhaps the only answer is a geographical auction in which the people bid for locations, taking the pits at salary levels which they feel compensate them adequately, possibly at the expense of those who take the peaks. After all, given finite budgets, the additional compensation for those who serve in the rural areas, for example, should come from those who stay with the bright lights of the city. It seems only fair.

6 Mental maps in today's world

'Map me no maps, sir, my head is a map,
a map of the whole world.'

Henry Fielding, *Rape upon rape*

As the ecological crisis grows, as distances shrink, we are finally becoming aware that we are all inhabitants of a single world – what Barbara Ward (1966) termed the 'Spaceship Earth'. Places continue to converge upon one another in both cost and time, and archaic national systems are seen to be but small components of a larger system vital to us all. The nation that dumps atomic wastes in the water, on the land or in the air, pollutes the globe, not simply the dumping grounds. DDT and other highly stable insecticides sprayed over the land eventually travel through biological amplification systems (food chains), and end up in the diets of animals and people. The Bermuda petrel, and many land and sea predators, will probably be extinct soon, since they can no longer lay hard-shelled eggs. Mercury levels are so high in many fish that they can no longer be consumed by people. Sulphur dioxide, and even more toxic industrial emissions, are not respectors of national boundaries. Huge sewers, that used to be called rivers, carry their filth to an international sea that is already locally polluted. The virtually closed Mediterranean can carry little more, while poisonous gases from World War II are seeping slowly over the floor of the southern Baltic. Only now, and still with agonising slowness, are we seeing ourselves in a much larger human context, realising that what we do, and how we live, may affect others today – and certainly our children tomorrow.

And our mental images of other places and other people are also slowly changing as we can no longer escape the reality of a closely connected world – a spider web of human relationships that transmits a quiver of distress or danger through the whole system. Distances that isolated, and yet buffered and protected us in former times, are becoming less effective. The permeability of geographical space is increasing, not simply for the diffusion and movement of goods and people, but for flows of communications and information. For some people in the nodes of highest communication intensity, often the swelling cities of megalopolis, the information streams and other stimuli are so intense and pounding that men and women can no longer cope with such a severely stressful man-made environment. Even as we tamper with delicate and little-understood physical systems, we are creating invisible landscapes of environmental stress that directly affect human behaviour and mental health. A pioneering study in the social psychiatry of Manhattan was laughed away by many as a gross exaggeration in the 1950s (Srole 1962). Today we are no longer laughing. We know that parts of our cities are peaks of psychic stress on the perception surfaces that describe collective images of geographical space.

IMPLICATIONS OF MENTAL MAPS

Throughout this book we have explored the mental images that people form of places, and we have looked at some of the things that may lie behind them. We have also touched upon the invisible landscapes and spaces that seem to play such an important role in forming the images of the world around us – the mental maps that seem to be shared to such a high degree, and which are so often predictable. Even though we still know very little about them, we would like to explore some of their implications for the future.

We are handicapped in making broad inferences directly from the evidence as almost all the respondents were literate – schoolchildren, school leavers or university students. However, we know from these examples that images of residential desirability are reasonably stable and predictable in the aggregate. We can make the plausible guess that broad patterns of migration are linked to the perceptions that people have of geographical space. And this is an important key.

Persistently high levels of rural–urban migration are the most outstanding feature of spatial reorganisation in the Third World today. The plummeting of expectations in the 1970s, as the worldwide recession has taken hold, has not changed this. For although the economic base of Third World cities has weakened in this period, the rural areas have suffered even more. As most central governments subsidise food prices to keep the lid on urban discontent, it is the small farmer who pays the price. In effect, the rural–urban terms of trade have been depressed by a heavy governmental hand.

So despite the worsening of the Third World's economic situation, the pull of the bright lights of the city is as strong as ever. This pull operates at all levels of the hierarchy, down to a level where the bright lights may be pretty dim by Western standards. In northeastern Nigeria, for example, the Rural Electrification Board puts a generator and a string of street lights into a small town. Almost immediately the corner gas station becomes the hangout for groups of rural youths – they have, perhaps, three years' education, no taste for farm life and are looking for somewhere to go.

Back in the United States, the low perceptual valleys and sinkholes over the Great Plains are all areas of out-migration, especially of the young and well educated groups, who are the most mobile in this very mobile society. The Southern Trough is also an area of high out-migration, although we must be very careful here, as the image of the South for some groups of people appears to be slowly changing. If we take a large group of university students from Pennsylvania as typical of today's young, well educated and upwardly mobile people, and compare their mental map with a similar group of 17 years ago (recall Fig. 3.3), we see some remarkable changes (Fig. 6.1). The entire southern portion of the USA is much more highly regarded than it was before, with the greatest changes in the 'Research Triangle' area of North Carolina, although clearly South Carolina and Georgia have also benefited greatly by these changing perceptions. There is also a distinct shift to the Rocky Mountain West (Montana, Wyoming, Colorado) and South-West (Arizona and New Mexico); an indication, perhaps, that the images of mountains, outdoor life and fewer constraints are exerting an even stronger drawing-power. In marked contrast, the Midwest and Great Plains are almost universally disparaged. These evaluations seem to be shared by their own young people, as these are some of the areas losing their populations – particularly the young, well educated adults.

The mental maps of today do resemble those of two decades ago, although they are nevertheless changing in consistent ways. The notion of the 'Sunbelt', while disparaged as a concept by many geographers, economists and other social scientists, is clearly a meaningful one for many young adults today who tend to 'vote with their feet'. In the 1960s the geographer William Warntz (1964, 1965) modelled the changing South as a human 'weather system', drawing a close analogy to an income 'front' that will move through the South by the year 2000, and integrate it finally with the rest of the country. So far he seems to be right.

There are other implications for the USA. By the year 2000, we may hope that the mental maps of Americans, northerners and southerners, will no longer reflect the wound established over a century ago. Even today this is kept open and sore in many formal and informal ways, ranging from established political characteristics, to teaching in the schools, to bumper stickers and flag waving at football games. Political reforms, giving proportional representation, will help the more liberal cities of the South, and from these increasingly dominant urban nodes changes will undoubtedly spread to the rural areas. Perhaps one day American schools – all American schools – may teach the Civil War as a tragic lesson in human compassion and understanding, rather than another lesson in Us against Them.

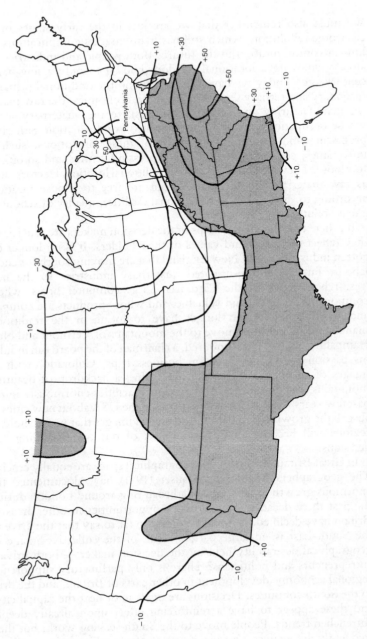

Figure 6.1 USA: population centres of the states.

MENTAL MAPS AND LOCATIONAL DECISIONS

We must also remember that we are just in the early stages of an electronic revolution, which still has many dramatic implications for human communication in the future. Both people and industries are already enjoying a locational freedom known to only a few half a century ago. Today we have moved beyond the traditional primary, secondary and tertiary sectors in advanced economies (the raw material, manufacturing and service industries), to the quaternary sector whose occupations are all deeply involved in information gathering, processing and dissemination. Most of these institutions, such as universities, consulting firms, data-processing centres and so on, are footloose. They are no longer tied to traditional locational factors, such as raw materials and the market, but are free to choose locations according to such things as recreational facilities, quality of schooling and a pollution-free atmosphere.

It is here that the mental maps of the decision makers are crucial, for they reflect the biases and values of their holders. It is no longer the cotton industries of old New England that are moving south for cheap labour, but the future-oriented quaternary industries to the new 'research triangles' of the Carolinas. Large computer firms, whose components are small and valuable, and whose products are compact and worth millions, no longer have to locate in the traditional manufacturing belt, but move to the mountains of Vermont and New Hampshire, where, it is rumoured, a chairman of the board can indulge his passion for skiing, between business trips. Colorado, with its images of an open and exciting life against a backdrop of beautiful mountains, has developed its recreational facilities enormously in the past few years. And as discussion grows in the USA about new cities to absorb the growing population, the mental images that people hold of regions will become an important part of our understanding and decisions.

In Great Britain the images of geographical space are equally crucial. The geographers Keeble and Hauser (1971) have documented the enormous growth in new jobs in a broad ring around London during the past three decades, even as work opportunities in other areas of Britain have declined. Is it really going too far to say that the drive to the South-East is, in large part, a result of the collective image of geographical desirability held by the decision makers – both private entrepreneurs and politicians? There is only perfunctory support for regional economic development in other parts of Britain, and the drift to the South continues. Decisions are made in or near the capital city, and these appear to have a reinforcing effect upon already deeply entrenched trends. People move to the South seeking work, but they frequently act against a strong desire to stay within their own local

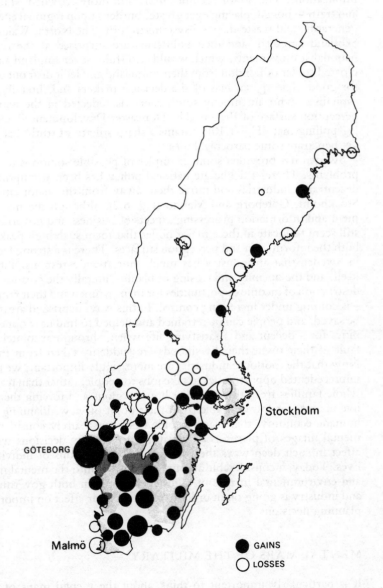

Stockholm

GÖTEBORG

Malmö

● GAINS
○ LOSSES

Figure 6.2 Changes in industrial employment in Sweden, 1960–65, showing marked
decentralisation from Stockholm and Malmö to smaller towns.

domes of desirability. We seem incapable of thinking through the implications. The South becomes more and more crowded, slummy and traffic-choked, placing ever greater burdens upon regional energy-generating and waste-disposal systems in turn. The North, Wales and Scotland languish, and then politicians are surprised at the rise of nationalist movements, which would cheerfully sever England and its bloated spider of London from their homeland and let it drift out to sea for good. The spatial bias of the decision makers and, literally, the capitalists, who are mostly southerners, is reflected in the national perception surface of Britain. The Doncaster Development Council's leg-pulling map (Fig. 1.10) contains a sharp stiletto of truth that may yet penetrate some parochial hides.

Sweden has provided some examples of possible solutions to such problems. There, a deliberate national policy has been attempting to decentralise industries and move them away from the major cities of Stockholm, Göteborg and Malmö (Fig. 6.2), although the management and information processing aspects of business and government still seem to locate in these major nodes that form such high peaks on both the information and perception surfaces. There is a strong feeling in Sweden that an industry has much more than a responsibility to itself, and the opening and closing of plants – literally the creation and destruction of income opportunities for men, women and their families – is coming under increasing control. In this way, depressed areas can be saved, and people can be retrained and helped to find new opportunities for a decent and a satisfying life when, through absolutely no fault of their own, their livelihoods are suddenly taken from them. Now that the footloose industries are increasingly important, we have unprecedented opportunities to take jobs to people, rather than tearing whole families from their familiar home regions and forcing them to live in areas far less desirable from their point of view. Planning in a humane economy, that places people first, must surely consider the mental images of places as a crucial input to policy decisions which affect in such deep ways the patterns and satisfactions of individual lives. Today, a considerable amount of research into the mental maps and environmental images of high-level people in both government and industry is going on in order to examine their effect on important planning decisions.

MENTAL MAPS OF THE MILITARY

It is particularly important to think about the mental maps of one group of decision makers. These are the images that military officers hold of the world around them, and they disclose some extraordinary evaluations. In a series of remarkable studies, Richard Eaton (1985), a

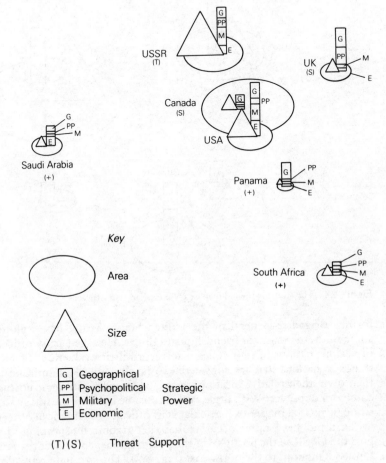

Figure 6.3 The strategic mental map of an American staff officer.

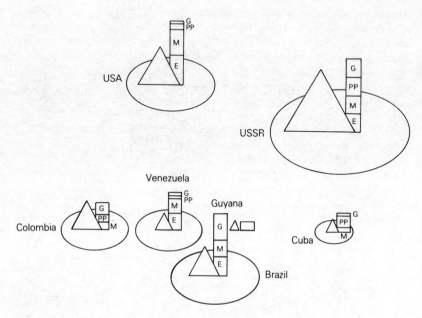

Figure 6.4 The strategic mental map of a Venezuelan staff officer.

former Brigadier-General of the United States Army, has explored some of these images of friendship and threat. He asked army officers of various nations in staff colleges to record their evaluations of such things as nations that are considered to be strategically significant to their own, their relative locations, sizes, populations, strategic posture and bases of power. With these evaluations, he built up the image, the strategic mental map, of a particular staff officer. An American officer, for example, has a rather accurate map, geographically speaking, that puts the USSR to the north, over the polar regions, and closer than the United Kingdom to the north-east (Fig. 6.3). The size and population images of Saudi Arabia and South Africa turn out to be quite incorrect, as are the relative populations of Canada and the UK. However, this strategic mental map sees the USSR as the only real threat (notice that China does not appear at all!), all other countries being seen as positively inclined or supportive. This is in marked contrast to a Venezuelan staff officer (Fig. 6.4) who has a highly contracted sense of strategic space in what appears to him a very hostile and threatening world. True, the USA and USSR are both prominent, with the psycho-political power threat more dominant for the latter, but he clearly has an almost paranoid view of all his neighbours, including tiny Guyana with its minute armed forces. The size and population of Brazil are grossly underestimated, and his perception of Cuba's threatening posture is mainly evaluated on psycho-political grounds.

The Venezuelan's rather threatened strategic view of his country is matched by an officer from Kenya (Fig. 6.5), whose strategic space is confined entirely to East Africa. In a sense, he lives in a world in which the USA, the USSR, China and Europe do not exist in any significant way for him as an army officer, although his evaluations of size, population and distance are remarkably close to reality. Only Ethiopia and the Sudan are friendly, and the greatest threat is seen as coming from Somalia. This tendency to 'contract' the strategic space to nearby countries also appears in the mental map of a Spanish staff officer (Fig. 6.6). Here the nations that are strategically important to Spain are located quite accurately, although he appears to be using an old-fashioned schoolroom Mercator projection, rather than the polar

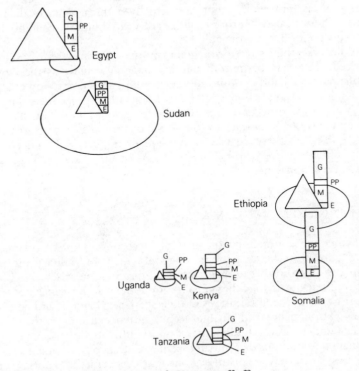

Figure 6.5   The strategic mental map of a Kenyan staff officer.

143

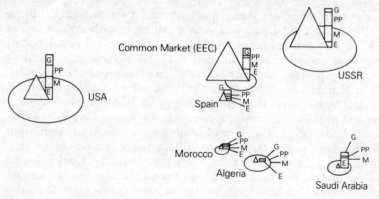

Figure 6.6 The strategic mental map of a Spanish staff officer.

projection of the American officer. Notice how the rest of Europe (excluding Portugal) has been lumped together in the 'Common Market', and its population grossly overestimated. The USSR is regarded as 'unfriendly', but not actually a threat – an evaluation reserved for nearby Morocco, with hardly any military, economic or psycho–political strength at all!

Of course, these strategic mental maps may not be extremely stable; indeed, they may be quite dynamic, changing rapidly as the world shifts politically and economically. General Eaton's point, however, is that it is important to understand how these rather special mental maps vary from the officers of one country to another, and from time to time. It may be important to follow the dictum of Robert Burns and 'see oursels as others see us', but it is equally important to understand how others see the world about them. After all, military officers are also decision makers.

CHANGING MENTAL MAPS

Whether we consider people making decisions to move, planners trying to create better patterns of employment or military officers thinking about threats to their countries, the question arises as to how we might try to change these images, or at the very least make people more aware of the implications of their spatial evaluations and perceptions. To influence and change deeply entrenched images that have been built up since childhood is a most difficult thing, sometimes demanding solutions that initially seem so outrageous, impractical and expensive that they are dismissed outright. Yet sometimes outrageous solutions of yesterday are tried, and we find tomorrow that they are really not so outrageous at all. Indeed, we often find ourselves accepting as perfectly obvious, rational and necessary a solution to a problem that seemed quite absurd to our more limited visions of the past. In Brazil, for example, the suggestion to move the capital city of Rio de Janeiro to a new site on the pioneer frontier of the interior was greeted as totally ridiculous. Yet it happened: and the simple awareness of the immense potential of the interior generated in the highly parochial legislators has already had startling effects. New trans-Amazonian highways are being built, and the dire predictions of links going from nowhere to nowhere have all fallen flat on their faces. Along the Brasilia–Belem trunk highway, that started as an earth road through the savanna and jungle, millions of head of cattle now graze, and new central places are agglomerating like growing beads on a string.

The pioneer fringes of Britain and the USA are admittedly somewhat different, but is it so outrageous and impracticable to think of holding Parliamentary and Congressional sessions in other centres, say Bristol, York and Inverness, and Atlanta, Denver and Seattle? Certainly the communication problem is no constraint when today we watch events directly half a world away. Men and women who make decisions about the national space and its people should be deeply aware of their own spatial and regional biases as they cast their legislative and essentially prescriptive and remedial votes, even as a psychoanalyst must be totally aware of his own personal psychic history, and the way it may affect his judgement of a patient. In brief, we all need a much greater geographical awareness of our own countries, and perhaps an even deeper understanding of others. Not the sterile geographies of yesterday – the can-you-name-the-state-capitals, and list-the-coalfields-of-Yorkshire varieties – but new and relevant regional geographies providing people with concept-illuminating information that helps them to see their Us–Them biases, their parochialisms, and the way they are linked to larger national and world communities.

And who are the people? Who are the ones we can realistically help at this point? They are, of course, the children – the one hope we all have of creating a more humane, less selfish, more understanding and compassionate world. Yet our educational and communication systems sometimes seem to be nothing more than institutions for the creation of human and geographical stereotypes. In China small children used to parade at school with toy rifles against the imperialists, even as children in the USA and Britain sit in front of TV sets and absorb an image of Africa that was never true. The mental maps of black children in Boston reflects Us and Them, even as the perception surfaces of Malay and Chinese students reflect deep, and here dangerous, cultural differences. In Northern Ireland stupid, inhumane and archaic religious sects divide children both in mental and geographical space, even as they divide the child of Quebec from his brother and sister in Ontario. On a larger scale the same sort of primitive, hatred–generating belief systems set one man to slaughter the children of another in Bangladesh. The children remain the one hope we have, yet we seem capable only of poisoning them as quickly as we can, generating in their minds images of other people, other places, and other ways of living that we have shaped and annealed in our vicious and savage adult world.

It is terribly sad to read that Trowbridge, the pioneer of mental maps, sought to justify his work on orientation with the observation that:

> ... the matter has a pertinent relation to the training of children to become soldiers, especially in countries where standing armies are maintained ... it would appear necessary that children should be seated at school in a special manner when studying geography, the cardinal points of the compass marked in the room, and the maps in the books properly oriented and the imaginary maps systematically corrected in childhood. (Trowbridge 1913)

Surely it would seem more constructive to use our knowledge of mental maps to reduce the need and the desire to go to war in the first place?

THE PERCEPTION OF EUROPE

In Europe the mental impress of the Iron Curtain, and other cultural and political divisions, can be clearly seen in the space preferences of people derived from a study carried out in 1965 to 1966 (Gould 1965). The most homogeneous group are the Swedes (Fig. 6.7), with a strong preference for Scandinavia, Switzerland, France and Britain. A sharp line separates eastern Europe from the rest, while countries with Catholic and authoritarian regimes are also viewed quite disparagingly. This view is almost identical to that from Britain ($r = 0.94$), with the exception of the views of Finland and Eire respectively. From Italy (Fig. 6.8) the same deep division appears, but there is a strong preference for a mixture of countries with Latin languages (France and Spain) and strongly democratic traditions (Britain and Sweden). The French hold somewhat similar views, but among them there is the least amount of agreement about residential desirability in Europe, and they seem quite divided in their opinions about the eastern countries. From West Germany (Fig. 6.9), the view appears strongly shaped by cultural and linguistic affinities, with Austria, Switzerland, The Netherlands,

Figure 6.7 Residential desirability in Europe as seen from Sweden.

Figure 6.8 Residential desirability in Europe as seen from Italy.

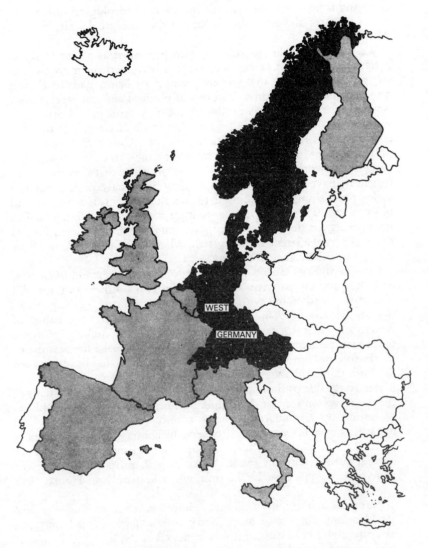

Figure 6.9 Residential desirability in Europe as seen from West Germany.

Denmark and Sweden all being highly perceived. The view of East Germany is in sharp contrast, for the effect of ideological differences places this area almost at the bottom of the perceptual scale, close to Albania.

We have raised the question of Europe almost as a case study in national division and perception, for it is here, perhaps more than in any part of the world, that serious attempts are being made by elected governments to overcome centuries of political and intellectual division in an area which has produced not only three cataclysmic local wars in the past century, but a long list of smaller conflicts in other parts of the world. The attempts to produce a United Europe are not only important for Europe itself, but offer hope – and danger – to other parts of the world. They offer hope, in the sense that if unity can be forged we have the first real example of international co-operation on a scale approaching that required to handle the sensitive human and physical world systems. But the attempts could also be dangerous if so much energy is devoted to European unity that Europe itself forgets that it is part of a larger world system, where many people do not have enough to sustain themselves.

Certain divisive aspects still appear to be very deep and tenacious, and the thought patterns of many Europeans still reflect the old boundaries marked out with such preciseness and care on the ground. Servan-Schreiber (1968), for example, has contrasted the habits of Americans and Europeans even within Europe itself, making the ironic point that it is Americans who seem to be able to make the best use of the European Common Market. At home Americans simply are not used to thinking about boundaries between states, and they are perfectly capable of dealing with somewhat different legal systems at the state level. In Europe their thinking simply overrides the problems of setting up headquarters in one country and branch plants and marketing systems in a dozen others. In contrast the Frenchman is hesitant about working in England, and the Englishman can hardly see himself dealing with the French – except perhaps in the very ancient wine-trade, going back to the time when England and France were one.

European unity, and world unity, imply much more than coal and Steel Communities, trade blocks and common markets. Such unity, in its deepest and best sense, implies a mental unity, a shared image of belonging to a much larger whole, even while retaining affection and respect for one's own home region and those of others. It is in the generation of respect and understanding that the educational and communication systems of Europe can do so much today for tomorrow. And it is here that the contemporary geographer has such an important role to play, not only in the schools and universities, but beyond these formal educational institutions into the wider public realm. Perhaps better than most scholars, but certainly not alone and without the help of many others, the geographer should be able to make us see ourselves as others see us; to break down the stifling parochialism, the boundary thinking, the Us–Themism; to create an awareness of what location in an information space implies for forming images and judgements. Only when we realise the way in which our collective perceptions are controlled and biased by our locations in streams of information can we begin to break out of the judgemental prison in which we are all trapped in greater or lesser degrees.

THE GEOGRAPHER AND EDUCATION

We have mentioned before the question of filters and filter control, for it is here that the geographer, with his particular spatial viewpoint, can push constantly for both educational and communication reform. In the schools, curriculum reforms in geographical education are long overdue, for many books to which young children are exposed are narrow, biased and dull. The first two are understandable; the last is unforgivable. Dull books, portraying mounds of facts without thought-provoking conceptual frameworks, do more harm than good, and it is hardly surprising that geography classes *per se* have been eliminated in the USA and Sweden. No child should be subjected to dullness, and if facts must be learned they should be taught as examples of more important general ideas. There are too many exciting and important concepts in the contemporary geographical viewpoint to eliminate them from the school curriculum so that children never become acquainted with them. It has been our experience, and the experience of many others in the USA and Britain, that young schoolchildren are fascinated by learning about other peoples and places, and many have an innocent curiosity about maps that can so easily be used as a platform for creative and imaginative thinking. At the older levels, deeper concepts and ideas can be discussed, as both John Cole in England, and the Geography High School Project in the USA, have demonstrated (Cole & Beynon 1982). At these levels, topics such as spatial diffusion, man–environment systems and urban growth dynamics can all form suitable geographical topics, increasing the awareness and sensitivity of the students to aspects of geographical space that can no longer be ignored by an informed electorate.

Universities, of course, are harder to move, but where new courses can be started in new settings we can see what can be done by imaginative men and women with clean blackboards to write upon. In the USA, the geography courses beginning at a few innovative universities, and at such new ones as the University of California at Santa Barbara, are very different from the dull and plodding intellectual pablums that mainly constitute feats of rote-memorisation. In Britain, the reform of the geographical curriculum has been remarkable over the past ten years, and many universities are now very strong in this area. The Open University continues to be one of the most exciting innovations in higher education this century, building upon the explosive developments of the past few decades by exposing first-year students to new concepts in an exciting, transdisciplinary adventure across all the social and behavioural sciences. Elsewhere in Europe geographical reform is moving more slowly, as the French geographer Jean Tricart (1969) documented for the Council of Europe. Sweden still has high standards, and many of its departments of geography continue their tradition of involvement with practical applied research. A few bright spots appear in The Netherlands, Belgium and West Germany, but much of French geography still seems in the doldrums. The University of Lisbon in Portugal has moved with great speed, joining a few younger faculties in Spanish universities, but elsewhere the geographical curricula of most European universities resemble those of 40 years ago.

The authoritarian, and increasingly bureaucratic, structure of the universities is partly to blame for this, since new ideas can be slowed down by powerful but conservative professors. But there is also the problem of language, one of the many barriers to free flows that can drastically filter a powerful flood of information down to a mere trickle. Despite our ability to communicate instantaneously across thousands of kilometres, we find ourselves transmitting only noise, not signal, if our next door neighbour does not understand us. Information spaces are complicated things, shaped not only by distances and arrangements of people, but by barriers of language, politics and religion. Our location in information space is important only if we can pick up the signals.

BARRIERS TO INFORMATION FLOWS

Yet it is precisely here that we have another task for the contemporary geographer. If Europe is to become united then we must break the barriers so that we can all share the same information space. That there are barriers to information flows in Europe and the world is a truism: but to measure and document these, and show the way in which potential acts of human interaction and communication are thwarted and often stopped completely, is to generate information and understanding that educated electorates, and sometimes even governments can comprehend. We know, for example, that communication decays with distance, that on the average the number of messages and the information per person decline with distance from the source. In a

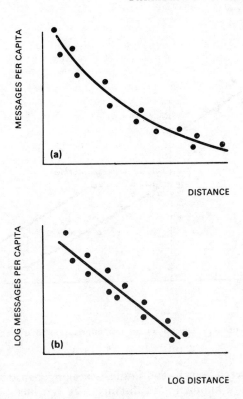

Figure 6.10 (a) Decay of information flows with distance plotted in standard co-ordinates. (b) Decay of information flows plotted in logarithmic co-ordinates.

homogeneous linguistic and cultural information space, we might expect a sample of information flows to plot in a typical decay curve (Fig. 6.10a) or as a straight line in logarithmic co-ordinates (Fig. 6.10b). But supposing some form of barrier exists in the information space – a mountain range, a long lake, or a change in language, religion or political system – what will our plot look like then? Obviously, just over the barrier the interaction will be considerably less than that expected simply from the usual distance–decay effect. We can describe the trend now with not just a single, best-fitting line, but two (Fig. 6.11). One will be fitted to the scatter of points on one side of the barrier, the second to the points on the other side. We now have a very simple way of estimating the effect of an information barrier. If we move our second trend to the right, we can make it line up once again with the first one describing only the normal distance decay. The distance we move it to the right becomes our measure of the barrier. For example, we might say that the linguistic and cultural barriers to

153

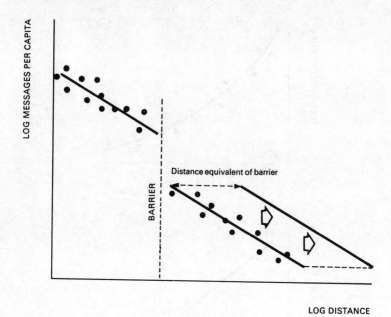

Figure 6.11 Measuring the effect of a barrier on information flows as an equivalent distance.

interaction between French- and English-speaking Canada are equivalent to so many kilometres of distance. We can literally think of French and English Canada being physically separated by this distance. This is precisely what the geographer Ross Mackay (1958) did in a study of information flows to and from Montreal, but his work could well guide a series of similar studies to document the effects of divisions in Europe at both the macro- and microscales. Think back a moment to the information fields and perception surfaces of the Swedish and Norwegian children, and the way in which even the highly permeable and open boundary between Sweden and Norway acted as a fault line, a break in the smoothly declining distance–decay effect (Fig. 6.12).

Yet we still know so little about these barrier effects, other than their simple existence. What, for example, are the relative effects of barriers between eastern and western Europe? Between France and Germany, as opposed to France and Belgium? Within Flemish and Walloon Belgium? In Switzerland? In Belfast and Londonderry? Between any group of human beings closed off from the larger information space in which they are embedded? You will recall the tiny urban world of the Spanish-speaking Mexican immigrants in Los Angeles (Fig. 1.7c). But what about large groups of southern Europeans in Swedish industrial cities, and the rich and poor in all places at all times – especially the

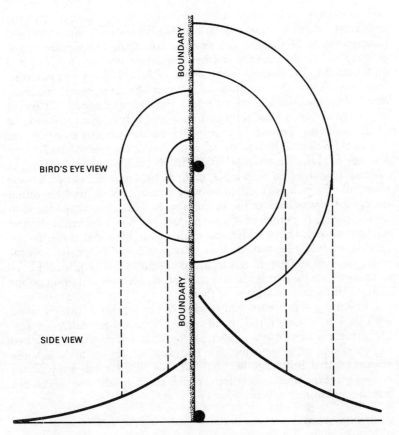

BOUNDARY

BIRD'S EYE VIEW

BOUNDARY

SIDE VIEW

Figure 6.12 A political, cultural or linguistic boundary forming a 'fault line' in an information field.

children? Notice, too, how we have descended a geographical scale, for barrier effects exist at all scales from the personal to the regional to the national and international. We know little about these effects at all these scales, and we have yet to start the difficult task of understanding how they all fit together in a nested hierarchy, in which regional effects are the collective result of many personal barriers which are, in turn, affected by the regional and national information flows and barriers themselves. Our personal images reflect our information flows, which are themselves shaped by the larger scales and barriers of the national space.

And so the geographer, with a deep concern for both the static geographic spaces of physical systems, and the relative spaces of information and perception, has many tasks ahead: tasks of teaching and research, two aspects of a life which can be separated only to the detriment of both, and tasks of public advocacy, education and understanding. In Europe, for example, the spatial viewpoint of the geographer would provide another important way of studying and understanding the massive complexity of both human and environmental systems. An Institute of Human Environmental Sciences devoted to the problem of a United Europe, and supported by all governments, could not only eliminate and help in the resolution of many problems, but set an example of macro-regional co-operation that could well be followed by other nations of the world. An African Institute of Human Environmental Sciences could result in a pooling of national resources for personnel, equipment, computers and physical plant, all of which are focused on very similar tasks of understanding the complex problems of life in the tropics. Linked, perhaps through UN agencies, to similar institutes in Asia and Latin America; sharing relevant knowledge quickly; and exchanging advanced students and faculties – especially students – such large and well supported institutions, devoted to focused research and advanced teaching, could have an immense impact on both the problems themselves and upon public and governmental awareness of them.

Our images – the maps and models of the world we carry around with us – need larger and much more relevant information inputs. Only then can our visions of a larger world, in which we are all linked together, and all responsible for one another, grow to match the human-created problems we shall *all* face shortly. For only if our visions meet the problems will the problems themselves be solved in a human, and humane, way.

Appendix: the construction of mental maps

When we talk about some degree of similarity between people's likes and dislikes for places, we imply that there is some relationship between the individual sets of preferences. Suppose we give Archibald and Hermione maps of their country which have been divided into familiar units such as the counties in Britain or states in the USA. We then ask them to rank these units in terms of their own personal preferences for where they would like to live. They could do this by placing a 1 in the unit they like best, a 2 in their second choice, and so on down to the area they like the least. Now it is extremely unlikely that their two sets of rank–order preferences would be exactly the same. On the other hand, it is quite likely that they would not disagree with each other completely: some relationship may exist between their two mental maps.

One of the ways we can see if there is any agreement between the evaluations of Archibald and Hermione is to plot their ranked preferences on a piece of graph paper. If their lists should happen to be exactly the same, the points representing the counties or states would lie on a straight line trending upwards and to the right (Fig. A.1a). There would be an exact relationship between their individual viewpoints, and we could measure this by calculating an index called a correlation coefficient. In this case of a perfect relationship, it would take the value +1.0. In fact Archibald and Hermione would probably not agree so completely. Instead of lying on a straight line, their preferences plotted on the graph might make a roughly oval scatter (Fig. A.1b), indicating some agreement between them, but not a perfect one by any means. In this case the correlation coefficient would be about +0.5. However, if the preference list of Archibald bore no relationship whatsoever to Hermione's, the scatter on the graph would show no overall trend (Fig. A.1c), and the correlation coefficient would be close to 0. It would, of course, be quite possible to have one other sort of relationship: the places Hermione tends to like Archibald dislikes, and vice versa. In this case the points would form a scatter on the graph trending downwards and to the right (Fig. A.1d), and the correlation coefficient would be about −0.5. The negative sign indicates an inverse, rather

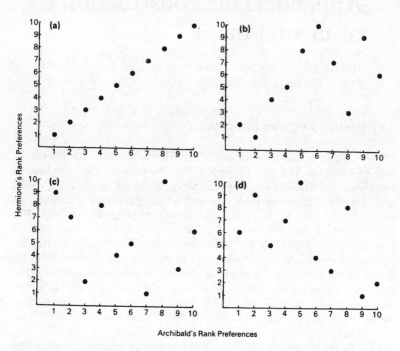

Figure A.1 (a) A preference plot by Archibald and Hermione: perfect agreement between them. (b) A preference plot by Archibald and Hermione: moderate direct agreement. (c) A preference plot by Archibald and Hermione: no agreement whatsoever. (d) A preference plot by Archibald and Hermione: moderate, but inverse agreement.

than direct relationship between their preferences. Thus a correlation coefficient, measuring the closeness of a relationship, can vary between +1.0, when there is a perfect, direct relationship; through 0, when there is no relationship at all; to −1.0, when there is again a perfect, but inverse relationship between two people's preferences.

When we ask people to rank their preferences for places, we are really asking them to measure the strength of their likes and dislikes on an ordinal scale – literally a scale that requires people to put things in order from the most- to the least-liked. We do not have to use this type of measurement, however, and we could ask them to assign to each unit on the map any value that they like between, say, 1 and 100 according to the intensity of their likes and dislikes for various places. If they did this, they would be measuring their preferences on an *interval* scale, because the intervals between the values they assigned would represent the differences between their evaluations of places. Alternatively, we could ask them to evaluate each county or state on a five- or seven-point scale, running from perhaps +3 for places they like very much, through 0 for places which they regard with shoulder-shrugging indifference, to −3 for places they dislike very much. They would then be measuring their preferences on a *nominal* scale.

As it turns out in practice, it does not make very much difference what scaling procedures are used, for they all lead to very consistent and similar results when we analyse the information people give us to construct their overall mental maps. Virtually all the maps we have discussed in this book are constructed from ordinal or rank data for two reasons: the first a rather practical one, and the second quite technical. When faced with the task of evaluating their preferences for places, many people find it rather difficult to assign interval measures, but feel it is relatively easy to rank places in order. They can take each area in turn and compare it to the others, saying to themselves, 'Now, if I really had to choose between these, which would I prefer?' Of course, not all ranks are equally easy to assign: most people feel it is very simple to rank the places they like and dislike very much, but there will nearly always be a number of areas in the middle to which they are indifferent, and the order of these may be a bit fuzzy. To borrow an analogy from electrical engineering, the order of the most- and least-liked areas are crisp signals, while the fuzzy, indifferent ranks in the middle may contain random noise. As we have seen, we need methods to pick out the crisp signals while filtering out the noise of uncertainty and indifference. The fact that this fuzzy area exists is not a problem as long as *some* crisp signals are emitted from either end of the scale. This point has not always been understood by some critics who have rejected the method because 'respondents cannot meaningfully rank 50 objects in order of preference'.

The second reason we prefer people to measure their preferences on an ordinal scale by ranking places from good to bad is that there is some very exciting but highly technical and mathematical work, which has shown that within a set of ordered objects there is, in fact, a great deal of information that the mathematician would term *metric*. Metric measurements are the ones with which most of us are familiar because they can always be shown as intervals between points. Fortunately, if we have ordinal data for a large group of people, it turns out that all this information very closely approximates the more precise metric measures of the interval scale. In fact, for any group of reasonable size, say over 20, the mental maps constructed from interval and ordinal measurements are virtually indistinguishable.

THE CONSTRUCTION OF A MENTAL MAP

Suppose a fairly large group of perhaps n people have all recorded their rank preferences for m regions on a series of maps. We can take this information from the individual maps and record it as a data matrix with m regional rows and n people (Fig. A.2). In this diagram only 7 people are recording their preferences for 10 regions – just to keep the example small and manageable. Each person, represented by a column of preferences in the matrix, will agree to a greater or lesser extent with all the others in the group, and we can measure their agreement by calculating the correlation coefficients between all possible pairs of people. These coefficients can be shown as a square 7×7 correlation matrix. Notice two things: first, that the terms along the diagonal are all 1.0, because somebody's ordered preferences are obviously correlated perfectly with themselves; and, secondly, that the matrix is symmetric – the relationship between Archibald (2) and Hermione (6) is obviously the same as between Hermione (6) and Archibald (2). Now let us convert the correlation matrix to a geometric diagram using vectors and cosines of angles. Each of the seven people, say Hermione, is a unit vector (Fig. A.3), and her agreement with all the others is indicated by the cosines of the angles made with all the other vectors representing the remaining $n-1$ or 6 people. We have seven people here, just to keep the diagram simple, and strictly speaking we should show the vectors in a seven-dimensional space, instead of just the two we are limited to by the pages of this book. Four or more dimensional spaces are too difficult to visualise anyway, so we must ask you to accept our word that this does not raise any mathematical difficulties when we come to a real problem.

Having arrived at a geometric picture of all the measures of agreement between people's viewpoints, let us recall our original task. We want to collapse all the individual preferences into one overall mental map by combining them in some way. The major question is: how do we combine all the individual values into one, overall viewpoint? Notice that our people–vectors have a tendency to thrust in one particular direction because there is some agreement between most of them. We shall try to measure this overall directional thrust of the vectors by passing a line or axis through their common origin (Fig. A.4), so that each vector makes a right-angled projection on this axis. Because the vectors are of unit length, their projections on the axis are nothing more than the cosines of the angles between the vectors and the major axis. Furthermore, as the big axis is moved around to point in the direction of the major thrust of the vectors, we might feel that it represents the most general overall viewpoint. Then, in this position, the cosines that each of the individual vectors make with this major axis

	PEOPLE						
	A	Archibald	C	D	E	Hermione	G
1	1	8	3	4	10	2	1
2	5	10	6	8	9	6	6
3	7	7	7	7	7	7	8
4	9	9	9	10	8	9	9
5	10	6	10	9	6	10	10
Areas 6	8	5	8	6	5	8	5
7	6	4	5	5	4	5	7
8	4	3	4	3	3	4	4
9	3	1	2	2	1	3	3
10	2	2	1	1	2	1	2

	A	Archibald	C	D	E	Hermione	G
A	1·00	·35	·95	·83	·20	·98	·93
Archibald	·35	1·00	·58	·79	·96	·49	·42
C	·95	·58	1·00	·93	·47	·99	·88
D	·83	·79	·93	1·00	·67	·90	·85
E	·20	·96	·47	·67	1·00	·36	·26
Hermione	·98	·49	·99	·90	·36	1·00	·90
G	·93	·42	·88	·85	·26	·90	1·00

Figure A.2 The basic preference data converted to a matrix of correlation coefficients.

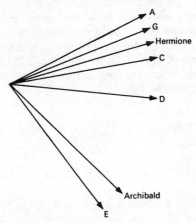

Figure A.3 The correlation coefficients shown as cosines of angles between seven vectors.

will measure the agreement, or correlation, of each person with the most typical, shared view.

The only questions to be resolved now are: How shall we locate this major, or principal, axis? What will be the criterion to judge whether it really does show the strongest directional thrust of all the vectors? Without going into a lot of mathematical detail, the criterion is a very simple one: we are going to swing the axis slowly around, and at each instant measure the lengths of the projections that the unit vectors, or

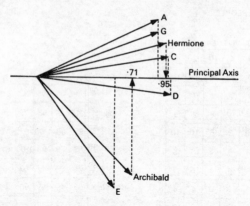

Figure A.4 A major axis through the vector cluster.

people, make on it. These projections, or cosines between the vectors and the axis, we shall call loadings. We shall say, for example, that a vector or person 'loads' on the principal axis, meaning that this value represents his or her correlation with the most general, shared viewpoint. What we want to do is find a position for the principal axis such that when we take each loading and square it, and then add all the squared loadings together, this quantity let us call it λ (lambda), is maximised. We might write the instructions out symbolically as follows: move the axis around until:

$$\sum_{i=1}^{n} 1_i^2 = \lambda \text{ is maximised}$$

In our example we have:

$$(0.90)^2 + (0.71)^2 + (0.97)^2 + (0.99)^2 + (0.59)^2 + (0.95)^2 + (0.89)^2 = 5.28$$

When we have found this position, we will stop and consider this the best axis for representing the major viewpoint of all the people in the group. Furthermore, since the loadings of each vector measure the correlation, or agreement, of each person with the most typical view, we shall use these as weights to combine the original rank values into overall scores for each of the regions.

This last step is just arithmetic: for example, if Archibald loads 0.71, then he agrees moderately with the overall viewpoint and his ranked values are weighted 0.71 in the overall score for each region (Fig. A.5). On the other hand, Hermione loads 0.95 because her viewpoint is related very strongly to the overall view of the group, so her ranked

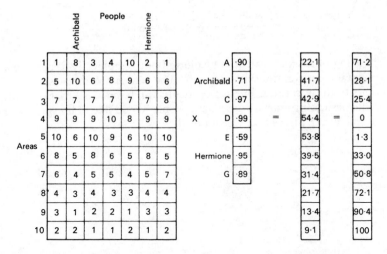

Figure A.5 Calculating the regional scores on the scale measuring the overall viewpoint.

values are weighted by 0.95. Thus, we calculate a score for each region by simply summing the rank values assigned to it by each person after these have been weighted by the respective loadings on the principal axis. For example, region 1 in our problem has a score of:

$$(1\times0.90)+(8\times0.71)+(3\times0.97)+(4\times0.99)+(10\times0.59)+(2\times0.95)+(1\times0.89)$$
$$=22.14$$

Because we let people rank their first choices as 1, 2, 3 and so on down to m (in this case 10), the best-liked regions will tend to have low scores, while those disliked will be high. Since this is inconvenient and not very sensible, we shall scale them so that the best-liked region has a score of 100, while the most disliked is 0. We can do this by taking the largest score away from each of the others, and then, ignoring the signs, divide through by the difference between the largest and smallest score. Finally, we multiply by 100 to give:

$$\frac{\text{scaled score}}{\text{region 6}} = \frac{\left|\begin{array}{c}\text{raw score} \\ \text{region 6}\end{array} - \begin{array}{c}\text{largest regional} \\ \text{score}\end{array}\right|}{\begin{array}{c}\text{largest regional} \\ \text{score}\end{array} - \begin{array}{c}\text{smallest regional} \\ \text{score}\end{array}} \times 100$$

The straight line brackets in the equation mean we take the absolute difference, ignoring the sign. In our numerical example we would have:

$$\frac{\text{scaled score}}{\text{region 6}} = \frac{|\,39.45 - 54.50\,|}{54.40 - 9.09} \times 100 = 33.0$$

These regional scores can now be plotted cartographically to construct a mental map for the whole group. The scores are used as control points to draw contour lines of residential preference, so that the hills on the desirability surface are places people would like to live in, while the valleys are places that are shunned.

Bibliography

Abelson, R. P. and J. W. Tukey 1959. Efficient conversion of non-metric information into metric information. *Proc. Social Statistics Section Am. Statistical Assoc.* 226–30.

Abler, R., J. Adams and P. Gould 1971. *Spatial organization: the geographer's view of the world.* Englewood Cliffs, NJ: Prentice–Hall.

Berry, B. 1970. Geography of the United States in the year 2000. *Trans. Inst. Br. Geogs.* **51,** 1–53.

Birmingham Post 1971. Buildings on a human scale in a confused city. 20 February.

Boal, F. W. 1969. Territoriality on the Shankill–Falls divide, Belfast. *Irish Geog.* **6,** 30–60.

Burris, F. 1970. *Patterns of ignorance of states' names.* Department of Geography, Northwestern University. Unpublished MS.

Canter, D. 1977. *The psychology of place.* London: Architectural Press.

Child, D. 1970. *The essentials of factor analysis.* New York: Holt, Rinehart & Winston.

Cole, J. and J. Beynon 1982. *New ways in geography,* volumes 1–4. Oxford: Basil Blackwell.

Curry, L. 1970. A spatial analysis of gravity flows. *Regional Studies* **6,** 131–47.

Davies, W. 1966. Latent migration potential and space preferences. *Prof. Geog.* **18,** 300–4.

Dornič, S. 1967. *Subjective distance and emotional involvement: a verification of the exponent invariance.* Reports from the Psychological Laboratories, no. 237. Stockholm: University of Stockholm.

Douglass, D. 1969. *Spatial preferences: a multivariate taxonomic and factor analytic approach.* MS thesis. Pennsylvania: University Park.

Downs, R. M. and D. Stea (eds) 1973. *Image and environment: cognitive mapping and spatial behavior.* Chicago: Aldine.

Downs, R. M. and D. Stea (eds) 1977. *Maps in minds: reflections as cognitive mapping.* New York: Harper & Row.

Eaton, R. 1985. Geopolitical mental maps: a search for synthesis. Ph.D. thesis. Pennsylvania: University Park.

Ekman, G. and O. Bratfisch 1965. Subjective distance and emotional involvement: a psychological mechanism. *Acta Psychologica* **24,** 446–53.

Expertgruppen för Regional Utredningsverksamhet 1970. *Urbanisering i Sverige: en geografisk samhällsanalvs.* Stockholm: Expertgruppen för Regional Utredningsverksamhet.

Forrester, J. 1971. *World dynamics.* Cambridge, Mass.: Wright-Allen.

Fotheringham, A. S. 1981. Spatial structure and distance decay parameters. *Ann. Assoc. Am. Geogs.* **71,** 425–36.

Gilluly, R. H. 1972. Into the abyss. *Intellectual Digest*, May, 24–5. (Reprinted from *Science News*, 1972.)

Goodey, B. 1968. *A pilot study of the geographical perception of North Dakota students*. Grand Forks: Department of Geography, University of North Dakota.

Goodey, B. 1971. City scene: an exploration into the image of central Birmingham as seen by area residents. *Research Memorandum* 10. Birmingham: Centre for Urban and Regional Studies.

Gould, P. 1965. *On mental maps*. Michigan Inter-university Community of Mathematical Geographers, Ann Arbor, Michigan.

Gould, P. 1967. On the geographic interpretation of eigenvalues. *Trans Inst. Br. Geogs*. **42**, 53–86.

Gould, P. 1969. The structure of space preferences in Tanzania. *Area* **4**, 29–35.

Gould, P. 1970a. Tanzania 1920–63: the spatial impress of the modernization process. *World Politics* **22**, 149–70.

Gould, P. 1970b. Var skall du vilja bo? *Geografiska Notiser*, 99–108.

Gould, P. 1972a. The black boxes of Jönköping. In *Cognitive mapping: images of spatial environment*, R. M. Downs and D. Stea (eds). Chicago: Aldine.

Gould, P. (1972b). A mental map of Ghana. *African Urban Notes* **6**, 4–5.

Gould, P. 1975. *People in information space*. Lund, Sweden: C. W. K. Gleerup.

Gould, P. and N. Lafond 1979. Mental maps and information surfaces in Quebec and Ontario. *Cahiers Géog. Québec* **23, 60,** 371–98.

Gould, P. and D. Ola 1970. The perception of residential desirability in the Western Region of Nigeria. *Environment and Planning* **2**, 73–87.

Gould, P. and R. White 1968. The mental maps of British school leavers. *Regional Studies* **2**, 161–82.

Hägerstrand, T. 1953. *Innovationsforloppet ur Korologisk Synpunkt*. Lund, Sweden: C.W.K. Gleerup.

Hall, E. 1969. *The hidden dimension*. Garden City, NY: Doubleday.

Harris, A. and F. Scala 1971. *Regional preferences and regional images in the United States*. Department of Geography, Northwestern University. Unpublished MS.

Harris, A., A. Patterson and R. White 1970. *Images of America*. Departments of Geography and Psychology, Northwestern University. Unpublished MS.

Heathcote, R. 1965. *Back of Bourke: a study of land approval and settlement in semi-arid Australia*. Melbourne: Melbourne University Press.

Johnstone, J., J. Willig and J. Spina 1971. *Young people's images of Canadian society*. Ottawa: Royal Commission on Bilingualism and Biculturalism.

Kaplan, S. 1972. Cognitive maps in perception and thought. In *Cognitive mapping: images of spatial environment*, R. M. Downs and D. Stea (eds). Chicago: Aldine.

Keeble, D. E. and D. P. Hauser 1971. Spatial analysis of manufacturing growth in outer south-east England 1960–67: hypotheses and variables. *Regional Studies* **5**, 229–62.

Ladd, F. 1967. A note on 'the world across the street'. *Harvard Graduate School Educ. Assoc. Bull.* **12**, 47–8.

Lee, T. 1963. Psychology and living space. *Trans. Bartlett Soc.* **2**, 9–36.

Leinbach, T. J. 1972. *Residential preferences in west Malaysia: general and ethnic surfaces.* Vermont: Burlington.

Ley, D. 1972. *The black inner city as a frontier outpost: images and behavior of a north Philadelphia neighborhood.* Ph.D. thesis. Pennsylvania: University Park.

Lundén, T. 1971. *Gränsskolan och det nationella systemet.* Kulturgeografiskt Seminarium, Kulturgeografiska Institutionen, Stockholm.

Lynch, K. 1960. *The image of the city.* Cambridge, Mass.: MIT Press.

Mackay, R. 1958. The interactance hypothesis and boundaries in Canada: a preliminary study. *Can. Geog.* **9,** 1–8.

Meadows, D. H., D. L. Meadows, J. Randers and W. Behrens 1972. *The limits to growth.* New York: Universe Books.

Meinig, D. 1962. *On the margins of the good Earth.* Chicago: Rand McNally.

Meinig, D. 1965. *The Great Columbia Plain: a historical geography, 1805–1910.* Seattle: University of Washington Press.

Michelson, W. 1977. *Environmental choice, human behavior and residential satisfaction.* New York: Oxford University Press.

New York Times 1971. A French view of New York: perilous city to visit. 24 January, 1.

Nystuen, J. 1967. Boundary shapes and boundary problems. *Peace Res. Soc. Papers* **8,** 107–28.

Orleans, P. 1967. Differential cognition of urban residents: effects of social scale on mapping. *Science, engineering and the city.* Publication 1498. Washington, DC: National Academy of Engineering.

Parkinson, C. Northcote 1967. Two nations. *The Economist,* 25 March, 1116–17.

Saarinen, T. 1966. *Perception of the drought hazard on the Great Plains.* Department of Geography research paper 106. Chicago: University of Chicago.

Servan-Schreiber, J.-J. 1968. *The American challenge.* New York: Atheneum.

Simon, H. 1969. *The sciences of the artificial.* Cambridge, Mass.: MIT Press.

Skinner, B. F. 1971. *Beyond freedom and dignity.* New York: Alfred Knopf.

Srole, L. 1962. *Mental health in the metropolis: the midtown Manhattan study.* New York: McGraw-Hill.

Stea, D. 1967. Reasons for our moving. *Landscape* **17,** 27–28.

Tricart, J. 1969. *The teaching of geography at university level.* London: Harrap.

Trowbridge, C. 1913. Fundamental methods of orientation and imaginary maps. *Science* **38,** 888–97.

Ward, B. 1966. *Spaceship Earth.* New York: Columbia University Press.

Warntz, W. 1964. A new map of the surface of population potentials for the United States, FIGO. *Geog. Rev.* **54,** 170–84.

Warntz, W. 1965. *Macrogeography and income fronts.* Philadelphia: Regional Science Inst.

White, R. 1967. *Space preferences and migration: a multidimensional analysis of the spatial preferences of school leavers.* MS thesis. Pennsylvania: University Park.

Wreford–Watson, J. 1969. The role of illusion in North American geography: a note on the geography of North American settlement. *Can. Geog.* **13,** 10–27.

Wreford–Watson, J. 1970–1. Image geography: the myth of America in the American scene. *Advance. Sci.* **27,** 1–9.

Wreford–Watson, J. 1971. Geography and image regions. *Geog. Helvetica* **1,** 31–3.

Index

man–environment relations 28
Mason–Dixon line 57
Massachusetts 57
Mbeya 125
measurement of mental maps 157–63
Megalopolis 8
Meinig, D. 26
metric measures 159
Metropolitan Sinkhole 42
Midland Mental Cirque 34, 40, 42
Midlands 8
Midlothian 40
migration 2, 31, 54, 129, 135–6
military images 140–44
Minnesota 61, 67
Minnesota's mental map 57
Mississippi 5, 54–9
model of perception surface 51
modernization surfaces 125
Monmouthshire 32
Montana 136
Montreal 96
Mormons 54
Morning Post, Apapa 123
Morogoro 125
Moshi 125
Mount Kilimanjaro 125
Mwanza 125

national or shared viewpoint 42
neighborhood 15–17
Nevada 54, 87
New England 57
New Hampshire 57, 138
New Jersey 61, 66, 67
New Mexico 5, 66, 136
New South Wales 26
New York 5, 57, 66
New York Times 108
New Yorker's map 17
Nigeria 115–17, 133
nominal scale 159
Norfolk 32
North Carolina 59, 67, 136
North Dakota 66
    mental map 61–3
    ignorance surface 83–7
Nyerere 124

Obuasi 127
Oda 127
Okitipupa 117
Okunnu, F. 132
Ontario 73–9
Open University 152

operant conditioning 28
ordinal scale 159
Oregon 54, 87
Orleans, P. 15–17
Oshawa 96
Oxfordshire 37
Oyo, mental maps 115–17

Paris 108
Parkinson, C. Northcote 23
Peace Corps (USA) 123
Pembrokeshire 32
Penang 130
Pennsylvania 57
    ignorance surface 83–7
    mental map 57
perception space 63–6
perceptual conflict 72
peripheral location 87, 90
Philadelphia 12
Pittsburg 8
planning 146
pollution 134
Pointe-aux-Trembles 81
Point Claire 96
population potential 6–11
Port Dixon 130
Portugal 3
Potter, B. 4
preference signals 114
principal axis 161–2
principal components analysis 30
psychic stress 12

Québec 5, 73–81
Queensland 26

rank preferences 157–61
Redcar's mental map 37, 51
relative location 119
residential desirability 4
Rhode Island 66
Rio de Janeiro 145
Royal Commission on Bilingualism
    and Biculturalism 76

San Francisco 27
Sansan 8
scales of evaluation 124
school leavers, maps of 32
scores, on principal component 162–4
semantic differential scales 124
Servan-Schreiber, J. J. 150
Sevenoaks' mental map 34, 44
Severn Bridge 32